Medical Tourism and Transnational Health Care

Medical Tourism and Transnational Health Care

Editor

Regina Belo

scitus
academics

Medical Tourism and Transnational Health Care
Edited by **Regina Belo**

ISBN: 978-1-68117-551-5

Library of Congress Control Number: 2016942958

SCITUS Academics LLC,
www.scitusacademics.com
Box No. 4766, 616 Corporate Way,
Suite 2, Valley Cottage,
NY 10989

Notice

Reasonable efforts have been made to publish reliable data and views articulated in the chapters are those of the individual contributors, and not necessarily those of the editors or publishers. Editors or publishers are not responsible for the accuracy of the information in the published chapters or consequences of their use. The publisher believes no responsibility for any damage or grievance to the persons or property arising out of the use of any materials, instructions, methods or thoughts in the book. The editors and the publisher have attempted to trace the copyright holders of all material reproduced in this publication and apologize to copyright holders if permission has not been obtained. If any copyright holder has not been acknowledged, please write to us so we may rectify.

Printed in United States of America on Acid Free Paper

Preface

Medical tourism refers to the travel of people to another country for the purpose of obtaining medical treatment in that country. Traditionally, people would travel from less-developed countries to major medical centers in highly developed countries for medical treatment that was unavailable in their own communities; the recent trend is for people to travel from developed countries to third-world countries for medical treatments because of cost consideration, though the traditional pattern still continues. Another reason for travel for medical treatment is that some treatments may not be legal in the home country, such as some fertility procedures.

The phenomenon of transnational health care has grown rapidly over recent years *Medical Tourism and Transnational Health Care* attempts to capture its implications for existing bodies of knowledge in selected aspects of medicine, medical ethics, health policy and management, and tourism studies.

As healthcare costs in the US and other parts of the world are excessively soaring, many employers and insurance companies started to view medical tourism as a way to lower them. More and more countries around the globe start to see the financial benefits from this emerging market, so they offer premium medical services at notably lower prices. The primary reason that clinics and hospitals in the developing countries are able to lower their prices is directly related to the nation's economic status. The direct correlation with per capita gross domestic product of the country is observed, which is a proxy for income levels. As a consequence, surgery prices are from 30% to 70% lower in the countries that are promoting medical tourism when compared to the US.The global growth in the flow of patients and health professionals as well as medical technology, capital funding and regulatory regimes across national borders has given rise to new patterns of consumption and production of healthcare services over recent decades.

Table of Contents

1

CHAPTER

Medical Tourism - Present Status & Challenges

Chennareddy Vijaya Mohanareddy[1], Chennareddy Rahulreddy[2]

[1] *Chief Executive Officer*

[2] *Trainee Observer Narayana Medical College Hospital, Chintareddypalem, Nellore - 524002, A.P., India*

ABSTRACT

Medical tourism or Health tourism is fast growing sector of medical marketing in INDIA and in many countries. The international health care market place emerged in the late 19th century when patients from less developed parts of the world began to travel to major medical centres in Europe and the United States to have diagnostic evaluation and treatment that was not available in their own countries. The situation is very different in the medical tourism model, where patients from highly developed nations travel to less developed countries, bypassing medical care that is offered in their own country but is inaccessible or more expensive to them. Another reason for travel for medical treatment is because some treatments may not be legal in the home country, such as some fertility procedures... People with rare genetic disorders may travel to another country where treatment of these conditions is better understood. Every type of health care, including psychiatry, alternative treatments, convalescent care and even burial services are available as a part of medical tourism.

Keywords: Medical Tourism, Health Tourism, Health-care Tourism

INTRODUCTION:

Medical tourism definitions:

1. 'A process of attracting foreign patients to overseas countries, which can offer Health care services at fees considerably less than the patient's home country and usually combining an element of post-operative tourism (recovery) for the patient'. (Rowley, 2008).[1]

2. Medical tourism defined as when consumers elect to travel across international borders with the intention of receiving some form of medical treatment. This treatment may span the full range of medical services, but most commonly includes dental care, cosmetic surgery, elective surgery, and fertility treatment. Medical tourism is an emerging global industry, with a range of key stakeholders with commercial interests including brokers, health care providers, insurance provision, website providers and conference and media services

Significance of medical tourism

With the advent of globalization, the world has shrunk into a global village. Countries now appear closer than they have ever been. The concept of 'medical tourism' was devised to help global citizens enjoy their tourist sojourns, while under cost-effective, world-class quality medical care. Over the last decade, India has emerged as the preferred destination for medical (healthcare) tourism, capitalizing on its strengths of "lower cost skilled personnel, cultural factors, natural endowments and unique forms of medicine."

Now a days Significances of medical tourism increasing popularity of medical travel include the high cost of health care, long wait times for certain procedures, the ease and affordability of international travel, and improvements in both technology and standards of care in many countries. The avoidance of waiting times is the leading factor for medical tourism from the UK. In the US, the main reason is cheaper prices abroad.

Many surgical procedures performed in medical tourism destinations cost a fraction of the price they do in the First World. For example a liver transplant that costs $300,000 in America costs, about $91,000 in Taiwan, where as in India it costs less than $50,000. . Countries that operate public health-care systems often have long waiting times for certain operations. Using Canada as an example, an estimated 782,936 Canadians spent time on medical waiting lists in 2005, waiting an average of 9.4 weeks. Canada has set waiting-time benchmarks, e.g. 26 weeks for a hip replacement and 16 weeks for cataract surgery, and for other elective surgeries.[2]

Aims & objectives

In this article an attempt is made to review the various aspects of Medical tourism with special reference to India, the Pros and Cons, the various measures to be taken, the difficulties faced, economic impact and others. The objectives are

- to study the various aspects of the Health care industry in India with respect to the Marketing segmentation, issues at hand and opportunities of growth.
- To cover over all global marketing sector of medical tourism
- To dwell on history and Process of Medical Tourism
- To discuss various Risks and Legal issues in Medical Tourism
- To discuss the current status and challenges of Medical Tourism in India and Global scenario

Origin and Growth of Medical Tourism

Rise in health care costs in developed countries, coupled with the availability of high quality medical services at a low cost in developing countries have given a boost to the medical tourism industry. These medical services range from elective procedures such as cosmetic surgeries to complicated surgeries such as cardiac, orthopaedics, neurosurgery and others. Significant growth in this industry is due to economic developments in developing countries that in turn have led to the growth in the medical industry and quality of medical services.

Medical Tourism Market (India, Thailand, Singapore, Malaysia, Mexico, Brazil, Taiwan, Turkey, South Korea, Costa Rica, Poland, Dubai and Philippines) - Global Industry Analysis, Size, Share, Growth, Trends and Forecast, 2013 - 2019"The global medical tourism market was valued at USD 10.5 billion in 2012 and is estimated to reach a market worth USD 32.5billion in 2019 at a CAGR (Compound Annual Growth Rate) of 17.9% from 2013 to 2019.[3]

Rise in the healthcare expenditure in developed countries coupled with in the growing elderly population has also contributed to the growth of the medical tourism across the globe. Economic crisis in the U.S. increased the number of uninsured population, consequently further triggering the growth of this market. Recently, there are approximately 50 million uninsured Americans that are willing to travel abroad for affordable and quality medical care. Globalization and improved communication technology act as a catalyst to boost the growth of this industry.

The development of medical tourism industry is based on several factors such as reduced cost of procedures, long waiting time and high demand for cosmetic surgeries. Medical travel is not only witnessed for intricate procedures but is also expanding due to growing demand, for cosmetic and dental procedures since these are usually not reimbursed under regular health insurance policies. Patients travelling abroad can save from 30% to 90% on a procedure, including their travel expenditures, as compared to the medical costs in their own countries.

Long waiting time is another factor which has substantially boosted the growth of this industry. Patients based in Canada and U.K travel to low cost destinations for their treatments in order to avoid long waiting periods in their own countries and receive timely access to serious ailments. Medical tourism industry is dominated by the Asian region that has captured the maximum share of the market. This region is highly competitive owing to the presence of technologically advanced medical specialties, less stringent government regulations and attractive locations. Thailand and India are recognized internationally for their high end medical services and receive patients from across the globe.

Singapore is renowned for its healthcare infrastructure and receives patients primarily for complex medical procedures. Thailand, India and Singapore accounted for approximately 60% of the total revenue of the Asian region in 2012. The Latin American countries such as Brazil and Mexico attract maximum number of patients from the U.S owing to the geographic proximity and cultural similarities. Similarly, growth in Turkey is witnessed on account of growing demand of cosmetic surgeries.

Paradigm shift in the market is seen owing to strong competition among the players of this industry. Malaysia is poised to have a significant growth in this market and is likely to emerge as a fastest growing country in terms of medical tourism with a CAGR of over 25% from 2013 to 2019. This advent in Malaysian medical tourism market is due to cost advantage over Thailand and Singapore. Moreover, rise in popularity of Malaysian region is attributed to many factors such as advanced healthcare infrastructure, highly skilled professionals, visa benefits and others. Additionally, robust government support for promotion of Malaysian medical tourism has propelled the growth in this nation. Increase in the government initiatives and growing inclination of private sector hospitals towards medical tourism is further supplanting the growth of this market.

Key healthcare providers in this industry include the following countries.

Costa Rica, Poland, Canada, Argentina, Brazil, Cuba, Thailand, Singapore, Malaysia, Mexico, Brazil, Turkey, South Korea, Taiwan, Dubai, Philippines, India, Sri lanka,Panama

Medical tourism in India- current scenario

Indian tourism industry significantly contributing to the nation's GDP, foreign exchange earnings and employment. People across world are travelling to India to obtain medical and surgical care, while at the same time visiting the tourist attractions of the country. Medical Tourism in India is growing at an exponential growth of 25%-30% annually. Present paper highlights the potential of Medical Tourism industry in India. [4]

Medical tourism market in India is estimated to be around US $333 million (in 2004), accounting for around 1, 50,000 patients mainly from US, Britain, Gulf, Africa and the South Asia Association for Regional Cooperation (SAARC) nations and growing at the rate of more than 25% p.a. Currently there are more than 2000 hotels in the A Grade cities which can provide an accommodation of approximately 1, 00,000 rooms, which is very well equipped to cater the growing needs of this segment.

For medical tourism to succeed a consortium of Tour operators, Hospitals and Hotels (if an integrated set up is not available) is a must. The purpose of these tour operators / agents is to liaise with the insurance companies and the private / public medical practitioners in the European and American markets, so as to ensure a steady stream of patients. In addition, the tour operators also help in explaining the problems that the patients may face when they visit India. (For e.g. the unhygienic surrounding, the non-availability or inadequate transport system etc.)

The cost of the tour operators can be as high as 15 to 20% of the total cost of the service offered. Thus for e.g., the cost of the tour operator could be 15 to 20% of say US $ 69200/- (Cost of a bone marrow transplant in India) or 15 to 20% of say US $69350/- (Cost for a Liver Transplant in India). Additionally, while the hospitals sell the medical services, the hotels sell India as a tourist destination using the ploy of selling "exotica".) or e.g. the traditional therapies and treatments of Kerala). Price is the major selling point. The cost of an open heart surgery could be around US$ 150000 in USA or US$ 70000 in UK, while the cost of the treatment in India could be as low as US$ 3000. India's share in the global medical tourism industry will reach around 5 per cent by 2015-end. The growth of medical tourism industry is marked globally, encompassing around 50 countries in all continents. Asia itself generates revenues in billions and consists of 12.7 per cent of the global market. Major healthcare players in India such as Apollo and Fortis have reported 10 per cent of their revenues from the medical tourism segment.

Indian Medical Tourism Conference and Alliance (IMTCA): is a non-profit international multidisciplinary organization that promotes excellence in research, treatment and provision of better and alternative healthcare, provides education and training, acts as a platform for doctors and international patients and creates awareness on global opportunities. 'Indian Medical Tourism Conference and Alliance' on November 2, 2012 conference at Hyderabad, Technopak study, reviewed that Indian market size of medical value travel would cross Rs 62,000 crores by 2020 and Rs 2 lakh crores by 2025 from Rs 4,500 crores in 2010.India could be hosting 2.4 million medical tourists by 2020, four-times the number it catered to in 2010, and this figure is projected to increase to 4.9 million tourists by 2025.The main reason for India's emergence as a preferred destination is the inherent advantage of its healthcare industry. Today Indian healthcare is perceived to be on par with global standards. Some of the top Indian hospitals and doctors have strong international reputation. But the most important factor that drives medical tourism to India is its low cost advantage. Majority of foreign patients visit India primarily to avail of "First World Service at Third World Cost".

Business World India and Indian Brand Equity Foundation:

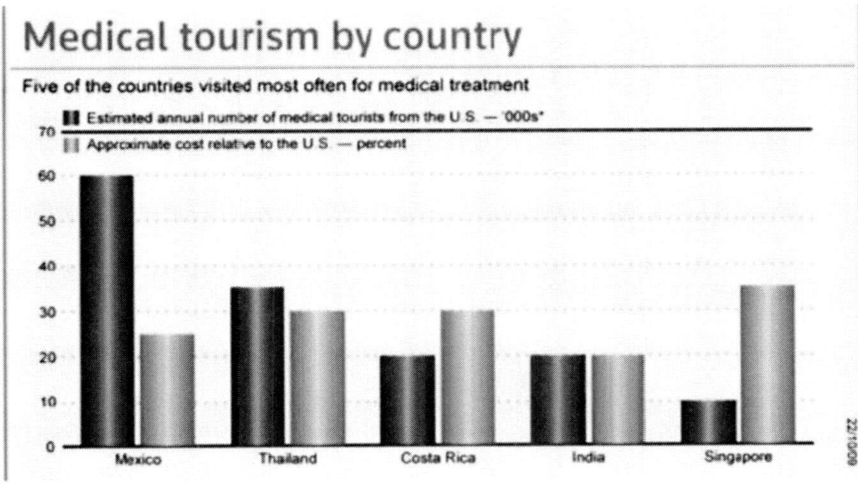

Table1. India has significant cost advantages in several health procedures making it a preferred destination.

Advantages & Challenges

The inflow of health tourists from the West, especially the UK, US and some of the European countries has been on the rise for the last couple of years. Price difference or affordability of the treatment, coupled with quality of doctors is the main reasons for the growing western traffic. The quality of Indian hospitals has improved significantly and now matches with the best in any part of the world. India has more than 100 healthcare institutions, which are of international standard.

According to Joint Commission Accreditation of Health Care Organisations (JCAMHO) the annual health bill of people from Afro-Asian countries seeking treatment outside their country is $10 billion. If India can tap even a fraction of that market, the potentials enormous.

Role of Government

- The role of Indian Government for success in medical tourism is two-fold:

- Acting as a Regulator to institute a uniform grading and accreditation system for hospitals to build consumers' trust. For facilitating investment, the policy recommendations include:

- Recognize healthcare as an infrastructure sector, and extend the benefits under sec 80- IA of the IT Act. Benefits include tax holidays for five years and concessional taxation for subsequent five years.

- The government should actively promote FDI in healthcare sector.

- Conducive fiscal policies - providing low interest rate loans, reducing import/excise duty for medical equipment.

- Facilitating clearances and certification like medical registration number, anti-pollution certificate etc.

- The above measures will kick-start hospital financing, which is struggling now due to capital intensive and low efficiency nature of healthcare business

For facilitating tourism the government should:

- Reduce hassles in visa process and institute visa-on-arrival for patients. Follow an OpenSky policy to increase inflow of flights into India.

- Create Medical Attachés to Indian embassies that promote health services to prospective Indian visitors.

Role of Private Sector

Increased participation in building infrastructure: To achieve its full potential, it is estimated that India needs an investment of Rs.100000 to 140000 Crores by 2012. Since the government can afford only a third of the amount, the private sector should play an active role to fill the gap.

Integrate horizontally: Private hospitals should also plan to integrate horizontally for providing end-to-end healthcare solutions to consumers. For example Apollo multi-specialty hospitals are already planning to set up Spas and alternative medi-clinics to attract more foreign tourists.

Joint Ventures / Alliances: To counter increasing competition, Indian hospitals should tie-up with foreign institutions for assured supply of medical tourists. Specifically tie-ups with capacity constrained hospitals and insurance providers will provide significant competitive advantage.

Innovation through medicities

Another successful example of the software industry is the establishment of Export Oriented Software Technology Parks. This model can be successfully replicated in the medical tourism industry by means of medicities. Each medicity could be a self-sustained healthcare hub with super specialty hospitals of international standards, ancillary facilities, research institutions, health resorts, rehabilitation centres and residential apartments. This model can be floated through a public-private partnership. The government will provide land and ancillary services and the private players will provide infrastructure and services. From the consumer's point of view, the medicities will offer superior value at affordable prices.

Health Care Services:

India offers world-class healthcare that costs substantially less than those in developed countries, using the same technology delivered by competent specialists and attaining comparable success rates. Indian medical education turns around 30000 doctors and nurses every year adding to the existing pool of over 14 lakhs doctors and nurses. About a dozen corporate hospitals provide world class treatments across all specialties. Availability of over 15,000 hospitals and 870,000 hospital beds provides adequate infrastructure support to the healthcare tourism. Also provide quality post-operative care with rehabilitation, naturopathy, physiotherapy, massage therapy and with best nursing care.

Alternative Medicine: With Ayurveda becoming increasingly popular across the globe, this ancient Indian therapy can be used as a non-surgical treatment for various ailments along with medication and yoga.

Hospitality:

In India hospitality is based on the principle "Atithi Devo Bhava", meaning "the guest is God". This principle is shown in a number of stories where a guest is literally a god who rewards the provider of hospitality. From this stems the Indian approach of graciousness towards guests at home, and in all social situations.

While coming to medical tourism aspect means proper accommodation and food to medical tourists. Know a day's hospitals and hotels take care of this service with pre and post-operative or procedure care. Sometime wellness clinics like rehabilitation centres will take care of services to the tourism. Rarely travel agency or tour organizers will take care of hospitality.

Travel services:

Services including flight reservation, welcoming service, airport transfer booking, transfer to preferential hotel booking and hospitals booking, hotel accommodation, arrival from home to departure.

Other skills and language: Refers to the ability to use hospital equipment's (i.e. utilizing clinical software, E-patient records, E-prescription, scheduling & reports, etc.) independently and converse fluently in English or the language required for the job.

Restrictive Government Policy:

India's large potential in medical tourism is being blunted by a thicket of restrictive rules for such visitors and lack of awareness campaigns overseas. Around 1.55 lakh such tourists came to India in 2010 but the number is seen as just a sliver of the Rs 4,000-crore global medical tourism market, says the study conducted by Indian Institute of Tourism and Travel Management and commissioned by the tourism ministry.

For India, the biggest problem is what the report dubs a "faulty" medical visapolicy". The rules allow only three entries a year, with a mandatory two-month gap between two entries. So an inbound patient, who travels for consultation with doctors, has to wait two months for treatment.

Another irritant is the requirement for such foreign visitors to report to government officials personally, a problem in cases of sickness. "The government should review the rule of personally reporting to FRROs (foreigners' regional registration offices) and devise a more realistic and humane way," [5]

Challenges before Indian medial tourism:

1. No strong government support or initiative to promote medical tourism.

2. Poor coordination between the various players in the industry- hospitals, air line Operators, and hotels.

3. Customer perception as an unhygienic country.

4. Lack of proper regulatory system for hospitals. 5. Lack of uniform pricing policies across hospitals.

5. Strong competition from countries like Thailand, Malaysia, and Singapore.

6. Lack of international accreditation.

7. Overseas medical care not covered by many insurance players.

8. Under-investment in health infrastructure.

Growth of Medical Tourism:

The countries where medical tourism is being actively promoted include Greece, South Africa, Jordan, India, Malaysia, Philippines and Singapore. In India, the Apollo group alone has so far treated 95,000 international patients, many of whom are of Indian origin. Apollo has been a forerunner in medical tourism in India and attracts patients from Southeast Asia, Africa, and the Middle East. The group has tied up with hospitals in Mauritius, Tanzania, Bangladesh and Yemen besides running a hospital in Sri Lanka, and managing a hospital in Dubai. Another corporate group running a chain of hospitals, Escorts, claims it has doubled its number of overseas patients - from 675 in 2000 to nearly 1,200 this year.

Recently, the Ruby Hospital in Kolkata signed a contract with the British insurance company, BUPA. The management hopes to get British patients from the queue in the National Health Services soon. Some estimates say that foreigners account for 10 to 12 per cent of all patients in top Mumbai hospitals despite roadblocks like poor aviation connectivity, poor road infrastructure and absence of uniform quality standards.

Analysts say that as many as 150,000 medical tourists came to India last year. However, the current market for medical tourism in India is mainly limited to patients from the Middle East and South Asian economies. Some claim that the industry would flourish even without Western medical tourists. Afro-Asian people spend as much as $20 billion a year on health care outside their countries – Nigerians alone spend an estimated $1 billion a year. Most of this money would be spent in Europe and America.

Promotion of Medical Tourism:

The key "selling points" of the medical tourism industry are its "cost effectiveness" and its combination with the attractions of tourism. The latter also uses the ploy of selling the "exotica" of the countries involved as well as the packaging of health care with traditional therapies and treatment methods. Price advantage is, of course, a major selling point. The slogan, thus is, "First World treatment' at Third World prices".

The cost differential across the board is huge: only a tenth and sometimes even a sixteenth of the cost in the West. Open-heart surgery could cost up to $70,000 in Britain and up to $150,000 in the US; in India's best hospitals it could cost between $3,000 and $10,000. Knee surgery (on both knees) costs 350,000 rupees ($7,700) in India; in Britain this costs £10,000 ($16,950), more than twice as much. Dental, eye and cosmetic surgeries in Western countries cost three to four times as much as in India.

The price advantage is however offset today for patients from the developed countries by concerns regarding standards, insurance coverage and other infrastructure. In India the strong tradition of traditional systems of healthcare in Kerala, for example. Kerala Ayurveda centers have been established at multiple locations in various metro cities, thus highlighting the advantages of Ayurveda in health management.

Problems Faced

1. India medical visa

Eligibility: One can get a medical visa, if you would like to seek medical treatment only in reputed/recognized specialized hospitals or treatment centres in India. Up to two attendants, who are blood relatives are allowed to accompany the applicant under separate medical attendant visas, their visa validity will be the same as the medical visa. Patient should carry medical certificate from local doctor. International Health Regulations (2005) norms must be satisfied for visa eligibility. [6]

Duration and Validity: The initial duration of the visa is up to a year or the period of the treatment, whichever is less. The visa will be valid for a maximum of 3 entries during the one year with minimum of 2 month gap up on previous visit. Visa duration starts on the day of issuance not from day of entry into India. [7]

Specific Rules: If you are travelling on a medical visa, you are required to register at the Foreigners Regional Registration Office within 14 days of your arrival. Patients from Pakistan and Bangladesh are required to register their arrival within 24 hours with the nearest Police Station/ Commissioner's Office. [7]

2. Ethics & legal issues

Ethical issues: There are many major ethical issues around medical tourism, example- the illegal purchase of organs and tissues for transplantations has been alleged in countries such as INDIA & CHINA prior to 2007.policy of medical tourism for the classes and health missions for the masses will lead to a deepening of the inequities. Medical tourism cantered on new technologies, such as stem cell treatments is often criticized as fraud and unproven therapies etc.

Legal issues: Receiving medical care abroad may subject medical tourists to unfamiliar legal issues. The limited nature of litigation in various countries is one reason for the lower cost of care overseas. Some countries currently presenting themselves as attractive medical tourism destinations provide some form of legal remedies for medical malpractice; these legal avenues may be unappealing to the medical tourist.

Hospitals or doctors in some countries may be unable to pay the financial damages awarded by a court to a patient. If problems arise, patient might not be recovered by adequate personal insurance or might be unable to seek compensation via malpractice lawsuits.

Many legal issues changes country to country, but high risk orientations, which may spread rapidly leading to higher causalities are banged in every country as per WHO- International Health Regulations 2005.

Prominent legal issues:

AIDS: HIV+:

India lifted all travel restrictions against HIVpositive patients in 2010.

In 2002, the Union health ministry had issued a notification, stating that mandatory test for HIV for India-bound foreign nationals should be removed from visa forms. However, some embassies failed to enforce it, and continued to display the requirement of HIV test certificate on their websites as well as on visa forms.

On September 17, 2010, the ministry of external affairs (MEA) clarified that there are no travel or residency restrictions for People Living with HIV (PLHIV) coming to India. [13]

HIV+ travellers can think of an Indian holiday. Currently there are no restrictions on entry to, or travel within, the country. The HIV status is not asked during the Visa application process or upon entry at Indian Customs.

Foreign students entering with an Indian Visa must submit an "HIV test report from one of the WHO recognized institutions from people in the age group of 15 to 60 years, if the foreigner is visiting India on a visa for more than one year". This report is one of many documents required for registration with the Foreigner's Registration Officer. This rule for Students is present round the world.(8)

Reproductive tourism and prenatal sex selection Abortion

Gender selective pregnancy is partial or complete illegal in 36 countries in the world – that have adopted National laws or policies having the force of law, on sex selection. Seven nations have already banned the use of gender selection technologies: Australia, Canada, China, India, New Zealand, Israel and the UK. 29 countries have stated that it opposes sex selection for non-medical reasons (medical reasons would be to prevent the passing on of genetic diseases). [9]

The World Health Organization (WHO), which has stated that sex selection for non-medical reasons "raises serious moral, legal, and social issues." Principal issues include, "the distortion of the natural sex ratio leading to a gender imbalance" and because it can "reinforce discriminatory and sexist stereotypes towards women by devaluing females"

The WHO defines sex selection as "the practice of using medical techniques to choose the sex of offspring, which "encompasses a number of practices including selecting embryos for transfer and implantation following IVF, and selectively terminating a pregnancy."

In nations where gender selection is banned, people often travel to the United States, Mexico, Italy, Thailand and other nations where it's legal to undergo PGD/PGS. This new phenomenon is called 'reproductive tourism' where people travel for gender selection and general infertility treatments such as IVF. For instance, in Australia, the Sydney IVF clinic discontinued its PGD sex selection program in 2005 after the Australian

Health Ethics Committee banned its use for nonmedical reasons. Many Australians now travel to Bangkok for sex selection procedures. Ironically, Superior ART in Bangkok operates in conjunction with Sydney IVF, its parent laboratory.

Fertility tourism or reproductive tourism is the practice of travelling to another country for fertility treatments. It may be regarded as a form of medical tourism. The main reasons for fertility tourism are legal prohibitions or regulation of the sought procedure in the home country, the non-availability of a procedure in the home country, as well as lower costs in the destination country. In-vitro fertilization and donor insemination are main procedures sought. It has been proposed to be termed reproductive exile to emphasise the difficulties and constraints faced by infertile patients, who are "forced" to travel globally for reproductive procedures.

Essentials of Health Tourism:

1. Knowing proper destination like country, hospital, how to travel.
2. Destination hospital or wellness centre approval and standards.
3. Travel to hospitals with proper visa like medical visa
4. Knowing current situation of states, where the tourist desires to travel.
5. Tourist should have the knowledge about states laws.
6. If travelling with insurance, see the countries and hospitals, which are approved by insurance companies.
7. Carry necessary documents and as much money as possible with some local currencies.

Health Tourism Agent: Many times customers are converted to medical tourists by Agencies. They guide the patient throughout the travel with a guide, if tourist demands. Agents will link up with travel agencies and hospitals on commission basis and provide service to customers, mostly with packages.

Medical Tourism Hospitals: Commonly provide package tothe overseas patient with guideline to travel safely to the hospital. Some hospitals tie ups with the Agents to get tourist. Before travelling to the tourist hospitals tourist must see the standards and rules of hospital as well as state.

Media: Internet advertisement, TV advertisement and newspaper advertisement media play a major role in motivating tourism to concerned hospital.

Accreditation: Concerns for the quality and safety of the medical care provided overseas have also emerged due to the lack of robust clinical governance arrangements and quality assurance procedures in provider organisations, intended to safeguard the quality of care provided to tourists . These bodies will organize the standards and train the processing to hospital in handling medical tourism. Initially accreditation organisation will provide some leads to its membership hospitals.

Treatment or Services: After health tourist landing in to overseas and hosted in hotel or hospitals for treatment purpose. Many countries provide high class hospitality to medical tourist. Major overseas Health services can be divided in to three patterns:

i. Wellness services

ii. Modern medical system

iii. Alternative medical system

After treatment: Patient will have their option to take rehabilitation program or return home. Patient may have the regular check up by local doctors or they will return to hospital where they took treatment or got surgery done for regular consultations. Nowadays patientsare preferring telemedicine after overseas treatment.

Telemedicine is the use of telecommunication and information technologies in order to provide clinical health care at a distance. It helps to eliminate distance barriers and can improve access to medical services that would often not be consistently available in distant communities. It is also used to save lives in critical care and emergency situations.

Early forms of telemedicine achieved with telephone and radio have been supplemented with video telephony, advanced diagnostic methods supported by distributed client-server applications, and additionally with telemedical devices to support in home care.

Common problems with medical tourism tourist faces:

i. Lack of research into the clinic or hospital.

ii. Lack of a proper consultation with the physician or dentist.

iii. Not considering their aftercare (for example, corrective treatment).

iv. Taking travel risks (for example, air travel after surgery).

v. Lack of insurance. Most insurers won't cover planned treatment abroad.

vi. Poor communication and language difficulties.

CONCLUSIONS

Medical Tourism in India is going to be more productive, cost effective and cost benefit for the patient, hospitals and nation. It's going to be good source of income to upcoming hospitals, if hospitals dedicate a separate international wing, to the patient services with infrastructure and good standards. India is going to be trusted as international medical tourist nation by standardization of hospitals with collaboration with international medical tourist leaders like JCI, ISQua and NABH. Comparative advantages of India are low cost, strong reputation in healthcare sector and diversity in medical system.

International insurance companies and agencies are playing a major role in converting patients to international medical tourist. It is felt that international insurance companies must be little more reliable towards overseas patients not only reimbursement of hospital services but also travelling and food expenses at least for a limit.

Various hospitals and health care providers across India are providing good health services to overseas patients. Major problem is that communication and memorandum of understanding, which plays a key role in guiding patients to hospitals. Even though language is a major barrier for the patients to communicate with hospitals, it is better that multispecialty hospitals should employ some multi language guides for international patient communication. The Medical Tourism Industry in

India is next big success story after software. According to a study the market size is estimated to be Rs.5000-10000 Crores by 2012. To realize the industry's full potential, a coordinated effort from the various players' government, private players and the associated sectors is very essential. The government should help in instituting an accreditation mechanism and device policies to facilitate private investment in the sector.

REFERENCES

1. Rowley, S. D. (2008,) Medical tourism in Asia. World Medical Tourism & Global Health Congress. San Francisco.
2. www .meditreatindia.com/ USA_medical_tourism_Services_surgeries.php...site
3. http://www.business-standard.com/article/ news- in Capital Market May 21, 2015 Last Updated at 12:00 IST on article Medical Travel Industry expected to grow at a CAGR of 17.9% from 2013-19 to reach USD 32.5 billion by 2019
4. INDIA TOURISM STATISTICS 2011 by government of India, Market Research Division and Ministry of Tourism.
5. http://www.indianmirror.com/tourism/ foreigners-registration.html 6. www.who.int/csr/ihr/WHA58-en.pdf
6. https://indianvisaonline.gov.in
7. http://timesofindia.indiatimes.com/india/ Travel-bar-on-HIV-patients-affecting-globaleconomy-UNAIDS/articleshow/ 17301168.cms..with title Travel bar on HIV patients affecting global economy: UNAIDS
8. http://timesofindia.indiatimes.com/india/ Government-sets-new-rule-for-sextest/ articleshow/45737879.cms

2

CHAPTER

Canadian Medical Tourism Companies That Have Exited the Marketplace: Content Analysis of Websites Used to Market Transnational Medical Travel

Leigh Turner

Center for Bioethics, University of Minnesota, Minneapolis, USA (55419), USA

ABSTRACT

Background

Medical tourism companies play an important role in promoting transnational medical travel for elective, out-of-pocket medical procedures. Though researchers are paying increasing attention to the global phenomenon of medical tourism, to date websites of medical tourism companies have received limited scrutiny. This article analyzes websites of Canadian medical tourism companies that advertised international healthcare but ultimately exited the marketplace. Using content analysis of company websites as an investigative tool, the article provides a detailed account of medical tourism companies that were based in Canada but no longer send clients to international health care facilities.

Methods

Internet searches, Google Alerts, searches on Google News Canada and ProQuest Newsstand, and searches of an Industry Canada database were used to locate medical tourism companies located in Canada. Once medical tourism companies were identified, the social science research method of content analysis was used to extract relevant information from company websites. Company websites were analyzed to determine: 1) where these businesses were based; 2) the destination countries and medical facilities that they promoted; 3) the health services they advertised; 4) core marketing messages; and 5) whether businesses marketed air travel, hotel accommodations, and holiday excursions in addition to medical procedures.

Results

In total, 25 medical tourism companies that were based in Canada are now defunct. Given that an estimated 18 medical tourism companies and 7 regional, cross-border medical travel facilitators now operate in Canada, it appears that approximately half of all identifiable medical tourism companies in Canada are no longer in business. 13 of the previously operational companies were based in Ontario, 7 were located in British Columbia, 4 were situated in Quebec, and 1 was based in Alberta. 14 companies marketed medical procedures within a single country, 9 businesses marketed health care at 2 or more destination nations, and 2 companies did not specify particular health care destinations. 22 companies operated as "generalist" businesses marketing many different types of medical procedures. 3 medical tourism companies marketed "specialist" services restricted to dental procedures or organ transplants. In general, medical tourism companies marketed health services on the basis of access to affordable, timely, and high-quality care. 16 businesses offered to make travel arrangements, 20 companies offered to book hotel reservations, and 17 medical tourism companies advertised holiday excursions.

Conclusions

This article provides a detailed empirical analysis of websites of medical tourism companies that were based in Canada but exited the marketplace and are now inoperative. The article identifies where these companies were located in Canada, what countries and health care facilities they selected as destination sites, the health services they advertised, how they marketed themselves in a competitive environment, and what travel-related services they promoted in addition to marketing health care. The paper reveals a fluid marketplace, with many medical tourism companies exiting this industry. In addition, by disclosing identities of companies, providing their websites, archiving these websites or print copies of websites for future studies, and analyzing content of medical tourism company websites, the article can serve as a useful resource for future studies. Citizens, health policy-makers, clinicians, and researchers can all benefit from increased insight into Canada's medical tourism industry.

Keywords: medical tourism Canada cross-border healthcare globalization websites

INTRODUCTION

The phrase "medical tourism" is often used to describe individuals travelling for health care and paying out-of-pocket for elective medical procedures [1, 2, 3, 4, 5, 6, 7, 8]. Medical tourism can involve travel within the borders of the country inhabited by a particular patient [9]. However, the term usually refers to transnational travel organized for the purpose of receiving medical care [10]. "Medical tourism" sometimes refers to regional health-related travel across borders dividing one country from another. For example, Canadians seeking health care and travelling from Ontario or Québec to Michigan or Vermont are occasionally described as medical tourists [11]. More commonly, the phrase is used to describe long-distance travel such as when a resident of Canada travels to India for medical care [12]. Individuals requiring emergency treatment seek immediate access to local hospitals and clinics. In contrast, medical tourism generally involves travel for elective, non-urgent medical interventions such as hip and knee replacements, dental procedures, and spinal surgery [13]. Patients might desire prompt access to these interventions but the treatments do not fall into the category of emergency care for life-threatening health conditions. Travel to spas, resorts, hot springs, and healing retreats is often characterized as "health tourism", "wellness tourism", and "spa tourism" [14, 15]. The phrase "medical tourism" is usually reserved for trips involving diagnostic tests and medical procedures falling within the scope of biomedicine. This distinction is not absolute. Many individuals combine different healing modalities and when travelling to international medical facilities seek both biomedical procedures and local healing traditions.

Travel for the purpose of obtaining medical care occurs for many different reasons and must be examined within the context of individual patient circumstances and large social-structural forces or political economies shaping access to health care [16]. For example, some uninsured residents of the U.S. travel to India and Mexico for medical care. Such individuals are often labeled medical tourists but other commentators, noting these travellers' lack of health insurance and inability to gain access to affordable care at local medical institutions, describe them as "medical refugees" or "medical exiles" [17].

Distinguishing among different types of health-related travel, researchers identify various kinds of medical tourists. Individuals travelling for kidney and liver transplants, for example, are sometimes given the label "transplant tourists" [18]. Patients seeking stem cell injections at clinics based in China, India, Mexico, Russia, and elsewhere are often called "stem cell tourists" [19]. The term "reproductive tourist" is used to describe individuals travelling for in vitro fertilization, other types of reproductive technologies, and commercial surrogacy [20]. Some scholars even use the term "suicide tourist" to describe individuals who travel to Switzerland for assistance in suicide [21]. Several of these labels are rather jarring and researchers disagree about the adequacy of various terms used to describe different types of cross-border medical travel. The primary objection to the medical tourism label is the linkage of medical procedures with the holiday-going, pleasure-seeking, leisure activities, and relaxation that the word "tourism" signifies to many individuals [22]. Though disagreement persists about what terms are best regarded as marketing labels and which phrases ought to be used by researchers, the phrase "medical tourism" is now routinely used in newspaper articles, policy documents, guidelines for patients, and peer-reviewed publications.

The subject of medical tourism attracts the attention of researchers from numerous disciplines. Scholars from anthropology, bioethics, cultural studies, law, public health, sociology, women's studies, and other fields have all made important contributions to the subject. Health law scholars provide insight into legal issues associated with cross-border medical travel [23]. Economists examine benefits and harms associated with increased international trade in health services [24, 25]. Bioethicists identify ethical issues and policy concerns they associate with medical tourism and the emergence of an inadequately regulated global marketplace in health services [26, 27]. In particular, they raise questions about quality of care and patient safety, continuity-of-care, commercialization of health care, public health considerations, health equity, and commodification of bodies of individuals selling organs or engaging in commercial surrogacy. Though previous research examining the phenomenon of medical tourism displays interest in how medical travel is situated in particular social, political, and economic contexts, many scholars note the need for more richly detailed, empirically-informed studies that address such topics as why patients travel for care, what types of procedures they undergo, where they travel, what quality of care they receive when they travel to international medical destinations, and what happens if they experience complications after receiving care abroad and then return to their local communities [28, 29].

BACKGROUND

The study of medical tourism companies, because of their intermediary role in facilitating medical travel, is an important component of the turn toward more empirically-informed and socially situated studies of medical tourism. To date, researchers studying medical tourism companies have paid particular attention to medical tourism companies based in the United States [30, 31, 32]. Journalists mention the existence of various medical tourism companies in Canada, and several articles in peer-reviewed publications identify different medical tourism companies in Canada, but there is limited scholarly research examining the medical tourism industry in Canada [33, 34, 35]. This article responds to a gap in scholarship and seeks to promote increased understanding of medical tourism in Canada. The text provides a detailed account of where defunct Canadian medical tourism companies were based, the destination sites they marketed to clients, the health services they promoted, and what they advertised in addition to medical procedures. This topic was addressed by compiling, between 2006 and 2011, a database of medical tourism companies based in Canada. The database contains companies marketing health services at international health care facilities and functioning as medical tourism agencies. It does not include websites that simply provide information about medical tourism. Once a database of medical tourism companies was established, content analysis of company websites was used to: 1) identify and document where in Canada particular medical tourism companies are based; 2) identify and summarize the destination countries to which these businesses indicate they send clients; 3) identify and record medical procedures, tests, and treatments marketed by these businesses; 4) identify and summarize the core marketing message of each of these companies; and 5) address whether medical tourism companies are involved in booking flights, arranging accommodations at hotels and resorts, and offering holidays at resorts, side trips, holiday excursions, visits to local landmarks and other activities often associated with the word, "tourism". The social science method of content analysis was used to address these five topics by analyzing websites of medical tourism companies. After content analysis was completed, medical tourism companies that currently market health care abroad were distinguished from defunct businesses. Medical tourism companies remaining in business are examined elsewhere. Companies that have ceased operations are the focus of this article. Unlike previous contributions to the study of transnational medical travel, the purpose of this article is to identify and analyze medical tourism companies that have exited the marketplace for transnational medical travel. In addition to its analytic function, the paper has historical value by capturing a turbulent period in the medical tourism marketplace and examining those Canadian businesses that are no longer involved in competing for clients. Medical tourism is often described as an industry with a rapidly expanding client base. This paper offers a more complicated account of the market for medical tourism by revealing that of all the medical tourism companies in Canada that could be identified over the last five years, approximately half of them are no longer in business.

METHODS

Development of a database of medical tourism companies located in Canada

Before analyzing websites of individual medical tourism companies it was first necessary to develop a database of all medical tourism companies with head offices or affiliate offices in Canada. Several methods were used to build this database. From 2006 to 2011, repeated Internet searches using Google Search Engine played a key role in locating medical tourism companies situated in Canada. Many businesses came into existence over the course of this study and were not identifiable during initial searches for medical tourism companies in Canada. Such phrases as "medical tourism company Canada", "medical tourism agency Canada", and "medical tourist Canada" were used as search terms. Searches for newspaper articles addressing such topics as "medical tourism", "medical tourism Canada", and "cross-border healthcare Canada" provided additional leads [36, 37, 38, 39, 40]. Newspaper articles were located using repeated searches on Google News Canada and ProQuest Newsstand. Newspaper reports of medical tourism companies situated in British Columbia, Alberta, Manitoba, Ontario, and Quebec played an important role in tracking the emergence of medical tourism companies across Canada. Google alerts for such terms as "medical tourism Canada", "stem cell tourism Canada", "transplant tourism Canada", and "global health care Canada" also generated extensive news reportage of Canadian medical tourism companies and Canadian patients travelling abroad for care. In addition, searches of an Industry Canada website that lists federally incorporated businesses in Canada yielded three company names. However, despite repeated efforts to locate these businesses it was not possible to find company websites, news media coverage concerning these businesses, or other signs indicating that these companies successfully entered the marketplace for promoting transnational medical travel. These companies are therefore not included in the final tally of twenty-five defunct companies with identifiable websites. Once company websites were identified they were bookmarked and added to the database. To facilitate content analysis, company websites were printed in their entirety. Content analysis was performed by analyzing printed versions of company websites and by periodically revisiting the Internet websites of medical tourism companies. Electronically archived websites or printouts of all identified medical tourism companies are available for review. Readers interested in assessing the content analysis provided in this article are welcome to access the primary data of the company websites.

Comparing the database to other lists of medical tourism companies in Canada

Internet searches resulted in the identification of several sources providing partial lists of medical tourism companies found in Canada. For example, in 2006, *Maclean's*, a popular news magazine in Canada, published an article that identified eight medical tourism companies in Canada [41]. Another list identifying five medical tourism businesses was found on a website promoting private health care alternatives and out-of-country health care options for Canadians [42]. A third source listed eight Canadian medical tourism companies within a longer record of medical tourism companies based in countries around the world [43]. Locating these lists permitted comparison of my database with other summaries of medical tourism companies based in Canada.

Content analysis of websites of medical tourism companies

After developing the database content analysis was used to study and analyze websites of all identified medical tourism companies based in Canada. Company websites were analyzed using pre-determined analytic categories; content was extracted for each category. The social science method of content analysis, as the phrase suggests, involves studying the content of such media as magazines, newspapers, films, television programs, and interview transcripts [44]. Content can be analyzed for general themes and concepts, images, particular statements, and other features. For the purpose of this study, medical tourism company websites were searched for specific pieces of information or website content. There were five main components to content analysis.

Location of medical tourism companies

First, I identified the city or town and province that medical tourism company websites identified as their business address within Canada. Identifying the location of medical tourism companies typically involved finding and then selecting such website tabs as "Contact us", "Contact information", or "Address" and recording information indicating where companies are located. Gathering this information resulted in a detailed record of where in Canada now-defunct medical tourism companies were based. In particular, it helped establish where they clustered. Though medical tourism companies in some respects are based "on" the internet, the physical location of companies matters for numerous reasons. For example, companies are governed by the legislation and regulations of the specific regions in which they are located. Within Canada, companies must operate according to both federal and provincial legislation; the province in which companies are situated informs which legal standards can be applied to them.

Destination countries and health care facilities

Second, I recorded the destination countries and/or destination health care facilities that now-inoperative medical tourism companies once marketed to their clients. Some medical tourism company websites identified particular hospitals and clinics. Other websites mentioned destination nations or networks of countries but did not list specific medical facilities. Websites were analyzed to develop a better understanding of destinations promoted by Canadian medical tourism companies.

Types of advertised medical procedures and medical specialties

Third, to provide insight into the types of health care packages medical tourism companies promoted, I recorded medical procedures or specialties marketed by medical tourism companies. Some companies provided lengthy lists of procedures. Other companies identified particular areas of medicine without listing specific procedures. Other businesses mentioned clinical specializations such as cardiology and gastroenterology or listed illnesses rather than focusing upon distinct medical procedures. This topic was explored to better understand what types of tests and treatments medical tourism companies marketed to prospective clients.

Core marketing messages of medical tourism companies

Fourth, websites of medical tourism companies were reviewed for statements that best encapsulated each company's core marketing message. These statements were typically located on the home page of company websites or in such sections as "About us", "What We Do", or "Mission statement". Core marketing messages were recorded and then summarized in brief point form. This topic was studied to develop a detailed account of the market niche now-defunct Canadian medical tourism companies attempted to occupy while they remained operational.

"Tourism" component of medical tourism

Fifth, company websites were reviewed for information about involvement of these businesses in booking air and/or ground transport, reserving hotel accommodations, and organizing tours, side trips, and/or other holiday excursions. This topic was studied in an attempt to determine whether holidays, side trips and other tourism-related activities were marketed in addition to health care packages.

In most instances, websites provided detailed information for each study question or category of content analysis. Data was recorded in tabular form. Where websites did not provide pertinent information, "NA" ("Not Addressed" in abbreviated form) was noted in tables.

The author of this article performed content analysis. The paper was fact-check three times by the author and feedback was solicited from two senior colleagues. All websites are disclosed and electronically archived. Readers interested in assessing the quality of content analysis have access to the primary data.

Distinguishing functioning from inoperative medical tourism companies

Following content analysis of all identified medical tourism company websites, in February and March 2011, internet searches, searches of an Industry Canada database [45] identifying the status of federally incorporated businesses in Canada, phone calls, and emails were used to distinguish active medical tourism businesses from medical tourism companies with expired websites, non-functioning email accounts, and discontinued phone numbers. Email and phone queries were used to determine whether businesses remained in operation. According to research ethics guidelines in both Canada, where this research project was initiated, and the United States, where it was completed, contacting a company to see if it remains in business does not fall within the scope of research requiring research ethics board review because there was no attempt to ask company owners or employees about their role within these companies, explore their attitudes toward medical tourism, investigate their workplace activities, or otherwise treat them as research subjects [46].

RESULTS

Using Internet searches, searches for newspaper articles mentioning medical tourism companies, and Google Alerts, it was possible to locate a total of twenty-five Canada-based medical tourism companies that have ceased sending their clients to hospitals and clinics located outside Canada. In contrast, operational medical tourism companies include an estimated 18 businesses marketing medical travel to such countries as Costa Rica, India, and Thailand and 7 businesses marketing regional, cross-border health services available in the United States as well as travel to private clinics within Canada. Several additional businesses are not comprehensive medical tourism companies but market bariatric surgery procedures performed in facilities based outside Canada as well as so-called "Liberation therapy" performed in India. Excluding these latter "boutique" businesses, approximately half of all medical tourism companies in Canada remain operational and the other half has disappeared from the marketplace.

Companies were labeled inoperative if websites were deactivated and phone lines were disconnected, company representatives reported that they were no longer accepting clients and had halted operations, they had lost their status as federally incorporated companies, or seven phone calls and/or emails failed to generate any response. It is possible that businesses that at present are not going concerns could at some point build a client base and begin sending customers to international medical facilities. If this occurs, in future analyses these companies will be reclassified as functioning medical tourism companies. My research provides a "snapshot" of Canada's medical tourism "landscape" at a particular moment; it is important to appreciate that the medical tourism industry in Canada is in flux. It is reasonable to assume that this industry is going to change over time as some companies emerge, other businesses disappear, and other companies transform themselves in response to client demand and perceived marketing opportunities.

Of the inoperative companies that I was able to locate, twenty-five had active, functioning websites at some point between 2006 and 2011. Additional file 1 identifies these businesses, provides website links, and contains Webcite references. Three additional companies, St. Luke Medical Tourism Center of North America Inc., Medical Tourism China, and International Medical Travel Corporation, were identified in the Government of Canada's database of businesses federally incorporated in Canada but do not appear to have had websites, issued press releases, or attracted any news media coverage. It is unclear whether these companies ever had clients even though they were incorporated businesses presumably established for the purpose of marketing medical tourism. These three businesses are noted but they are not included in the total tally of twenty-five defunct medical tourism companies. With no websites to analyze, and no indications that they had ever sent clients to international destinations, there was no basis for including them in the content analysis process.

Locations of medical tourism companies

Of the twenty-five companies that established websites and were studied using the method of content analysis, thirteen were located in Ontario, seven were based in British Columbia, four were situated in Quebec, and one was located in Alberta. Websites of some Canadian medical tourism companies identify affiliate offices or company representatives located outside Canada. Where such information was provided I recorded where company representatives based outside Canada were situated.

Table 1 lists company names, identifies where in Canada these businesses were located, and documents affiliates and representatives in those instances where companies had offices or agents situated outside Canada.

Table 1. Locations of Medical Tourism Companies in Canada

Company	City & Province
Axiom Health Solutions	Canadian Office: Oakville, Ontario; U.S. Office: Kensigton [sic], Maryland
Canadian Healthcare International (CHI)	Markham, Ontario
CubaMedicare	Oakville, Ontario
EcuMedical Resources International Ltd.	Windsor, Ontario
First Choice Medical Tourism	Canadian Office: Calgary, Alberta; with contact person in Luzon, Philippines
Health Trips	Ottawa, Ontario
Health Vacations, Inc.	Ottawa, Ontario
International Medical Network	Oakville, Ontario
JD Healthcare	Nanaimo, British Columbia
LAM International (L.A.M. Logistic Assistance Medical International)	Canadian Office: Toronto, Ontario; and Colombia
MedAsia	Montreal, Quebec
MedExpress Tourism	Montreal, Quebec
Medi-Pro Medical Management	Windsor, Ontario
MedSolution	Vancouver, British Columbia
Medtourlink	Vancouver, British Columbia
Reach Health Services & Outsourcing	Canadian Office: Vancouver, British Columbia; Representative in Chennai, India
Recover Discover Healthcare	Canadian Office: Vancouver British, Columbia; U.S. Office: Austin, Texas; India Office: Noida, India
Royal Med Services	Mississauga (Toronto), Ontario
Speedy Surgery Global Healthcare	Quebec City, Quebec
Star Hospitals	Canadian marketing office in Toronto, call centers and operations in Chennai, India
Sun Medical Group	Vancouver, British Columbia
The InciDental Tourist	Ottawa, Ontario
Tooth Tourism	Surrey, British Columbia
Unbelievable India	Dollard-Des-Ormeaux, Quebec
Victus Global Healthcare	Main Office: Toronto, Ontario; Additional office: Edmonton, Alberta

Destination countries and health care facilities

Of the twenty-five companies, fourteen marketed medical procedures at just one medical facility or at several medical facilities within one country. Three companies marketed health care in two destination nations, two companies advertised procedures in three countries, two companies promoted health services in four countries, one company marketed procedures in five countries, and one of the dental tourism companies marketed dental procedures in seven countries. Two companies did not provide details concerning where they sent their clients.

The twenty-five medical tourism companies marketed health services in twenty-one different nations. Twelve companies marketed medical travel to India; four listed Thailand as a destination site; three companies listed Singapore; Canada, Cuba, the United States, the Philippines, Costa Rica, El Salvador, China, Malaysia, and Mexico were all twice listed as destination sites; and Morocco, Russia, Hungary, Colombia, Taiwan, France, Turkey, Dominican Republic, and Panama were all listed once. Listing of countries as destinations does not mean that any residents of Canada ever selected these particular destinations as sites for medical care. However, documentation of advertised health care destinations provides insight into the partnerships or networks medical tourism companies choose to market to their prospective client base. While they do not provide information about actual flows of patients across national borders, they reveal how medical tourism companies that were based in Canada promoted health facilities in other nations. Table 2 identifies locations of destination facilities marketed on company websites.

Types of advertised medical procedures

Twenty-two companies marketed many different kinds of medical procedures and can be characterized as "generalist" medical tourism companies. They provide lengthy lists of different types of medical procedures. Three businesses marketed specialized services and restricted themselves to a niche position within the medical tourism industry. Of these latter businesses, two companies limited themselves to marketing dental procedures and one company advertised organ transplants in Colombia. Table 3 summarizes health services marketed by medical tourism companies.

Core marketing messages of medical tourism companies

Though there were variations in core marketing messages of medical tourism companies, most businesses emphasized affordability of care, timely access to medical care, and high-quality care. In total, of twenty-five businesses, twenty companies marketed access to affordable care, seventeen businesses advertized timely access to care, and twenty-one websites emphasized high-quality care. The few companies that did not mention all three features offered a subset of these offerings.

Table 2. Medical Tourism Companies and Destination Facilities

Company	Destination Facilities
Axiom Health Solutions	Apollo Hospitals, Wockhardt Hospitals, Escorts Heart Institute and Research Centre, Max Healthcare, Fortis Healthcare, Sahaj Dental Clinic, all in India
Canadian Healthcare International (CHI)	Health care facilities in Canada (particular institutions are not identified)
CubaMedicare	Medical facilities in Havana, Varadero, and Holguin, all in Cuba
EcuMedical Resources International	Health care facilities in the United States
First Choice Medical Tourism	St. Luke's Medical Center, Quezon City, Philippines; Estetico Manila, Manila, Philippines
Health Trips	German Collaborative Private International Hospital, India
Health Vacations, Inc.	St. Petersburg, Russia
International Medical Network	Hospitals and clinics in India, Hungary, Cuba, and Costa Rica
JD Healthcare	Website claims company has partnerships with more than 30 hospitals (particular institutions are not specified)
LAM International (Logistic Assistance Medical International)	Transplant facilities in Colombia
MedAsia	Min-Sheng General Hospital, Taiwan
MedExpress Tourism	Health care facilities in Morocco
Medi-Pro Medical Management	Laser Spine Institute, Tampa, Florida; Crittenton Hospital, Rochester, MI, Internal Medicine, Mount Clemens, MI; additional facilities in the U.S.
MedSolution	India, France, Turkey, El Salvador; website indicates plan to include Costa Rica, Malaysia, Mexico, Poland, and South Africa
Medtourlink	Singapore, Philippines, India, China, Canada
Reach Health Services & Outsourcing	Apollo Hospitals, Escorts Heart Institute and Research Centre, Rajan Eye Care, JGHR Dental, all in India
Recover Discover Healthcare	Escorts Heart Institute & Research Centre, Delhi; Fortis Hospital, Noida; Fortis Flt. Lt Rajan Dhall Hospital, Delhi; Fortis Hospital, Mohali, Chandigarh; Fortis-Wockhardt Hospital, Bangalore; Fortis-Wockhardt Hospital, Mumbai, India; Gleneagles Medical Center, Penang, Malaysia
Royal Med Services	Medical facilities in India
Speedy Surgery Global Healthcare	Medical facilities in India and Thailand
Star Hospitals	India, Singapore, Thailand
Sun Medical Group	Malaysia, Singapore, Thailand
The InciDental Tourist	Nanjing,China; Merida, Mexico
Tooth Tourism	Costa Rica, Dominican Republic, El Salvador, India, Mexico, Panama, Bangkok, Thailand
Unbelievable India	Medical facilities in India
Victus Global Healthcare	JCI-accredited medical facilities around the world (particular institutions are not specified)

Table 4 provides summaries of medical tourism companies' core marketing messages. Additional file 2 provides summaries of core marketing messages in addition to extended excerpts from company websites. These lengthy excerpts are intended to demonstrate the empirical basis for identifying and summarizing core marketing messages.

"Tourism" component of medical tourism

Core marketing messages rarely emphasized the possibility of patients undergoing medical procedures while also enjoying holiday excursions. However, activities commonly associated with the concept of tourism were marketed by more than half of the companies. Sixteen medical tourism companies offered to arrange air travel and/or organize ground transport. Twenty companies marketed the service of booking hotel

Table 3. Marketed Procedures

Company	Health Services Marketed
Axiom Health Solutions	Interventional cardiology, cardiothoracic surgery, cosmetic and plastic surgery, hand and micro surgery, medical & surgical gastroenterology, neurology & neurosurgery, spine surgery, surgical oncology, orthopedic surgery, vascular surgery, brain surgery, bone and joint surgery, eye surgery, minimal access surgery, obesity surgery, maxillofacial surgery, thoracic surgery, orthopedic surgery, dental care and dental surgery, transplant surgery, and other treatments
Canadian Healthcare International (CHI)	Specific procedures are not identified though total hip replacement and arthroscopic rotator cuff repair are used to compare cost of care in Canadian facilities versus institutions in other countries
CubaMedicare	Drug and alcohol rehabilitation, pigmentary retinosis, neurological rehabilitation, vitiligo, psoriasis, and alopecia, cosmetic surgery, odontology, dental implants, orthognatic surgery; alternative therapies, stroke rehabilitation, spinal cord injury rehabilitation, brain injury rehabilitation, physical, occupational, and speech therapy, aquatic therapy, physical therapy, anorexia nervosa treatments
EcuMedical Resources International Ltd	cancer, cardiology, cosmetic surgery, day surgery, diagnostic imaging, hip surgery, knee surgery, minimally invasive surgery, neck and spine surgery, neurology, orthopaedics, second opinions, urology, VIP services
First Choice Medical Tourism	General medical surgery, dental surgery, cosmetic surgery, optical surgery, spa retreats
Health Trips	knee replacement, hip replacement, shoulder replacement, spinal surgery, correction of deformities, beating heart surgery, cosmetic surgery
Health Vacations, Inc.	MRI outsourcing, endocrinology surgery, orthopedic packages, ophthalmology packages, spinal surgery, diagnosis and treatment of allergies, preventive health check ups, executive health check ups, preventive heart checkup, whole body check up, and additional treatments, access to specialists in cardiology, internal medicine, neurology, obstetrics/gynecology, orthopedic surgery, pediatrics, urology, endocrinology, rheumatology
International Medical Network	Cardiovascular, orthopedic, neurosurgery, female reproductive system, gastro-intestinal, cosmetic, dental, IVF
JD Healthcare	Assisted reproduction, plastic surgery, bone and joint surgeries, cancer treatment, cardiovascular surgery, eye surgery, general surgery, cerebrovascular surgery, dental surgery, dermatology, and additional treatments
LAM International (Logistic, Assistance, Medical International)	Heart transplants, liver transplants, lung transplants, kidney transplants
MedAsia	Bariatric surgery, cardiovascular surgery including minimally invasive cardiovascular surgery, PTCA and stent, and cardiac catheterization, orthopedic surgery including total joint replacement, spine surgery, and arthroscopy, other treatments
MedExpress Tourism	Dental implants, weight loss, cosmetic surgery, orthopedics, urology, cardiology, eye corrective (Lasik), cancer, general surgery
Medi-Pro Medical Management	MRI scans, CT scans, body imaging, X-rays, arthroscopy, bariatric surgery, cardiology, endocrinology, internal medicine, neurology
MedSolution	Cardiac care, chemical dependency programs, cosmetic & plastic surgery, dental care, gastric surgery, heart surgery, infertility treatments, orthopedic surgery, urogenital surgery
Medtourlink	Hysterectomy, cataract, Lasik eye care procedures, breast lift, tummy tuck, facelift, rhinoplasty, dental implants, dentures, crowns and aesthetic dentistry, MRI, X-ray, PET, CT diagnostic scans, laser procedures
Reach Health Services & Outsourcing	Cardiac surgery and cardiology, minimal access surgery and urology, orthopedics, ophthalmology, gastroenterology, obstetric and gynecology, oncology, cosmetic/plastic surgery, preventive health checks, diagnostic services
Recover Discover Healthcare	Knee replacement surgery, hip replacement surgery, hip resurfacing, adult & pediatric cardiac care, oncology treatments, bariatric care, laparoscopic gastric band, gastric bypass surgery, sleeve gastrectomy, neurosurgeries, spinal care, infertility management including IVF, cosmetic care, additional surgeries
Royal Med Services	Hip replacements, other joint replacements, breast augmentation, liposuction, and other procedures
Speedy Surgery Global Healthcare	Allergy, cardiology, dentistry, diabetes and endocrinology, ear, nose and throat diseases, eye laser refraction, oncology, ophthalmology, orthopedic, plastic surgery, preventive health check, radiology and imaging services, and additional procedures
Star Hospitals	cardiology, cardiothoracic surgery, orthopedic, neurology and neurosurgery, eye care, cosmetic surgery, dentistry, comprehensive and preventive health checks, weight loss, obstetrics and gynecology, IVF, additional treatments, yoga, Ayurvedic consultation, stress management, weight reduction, spine and joint care
Sun Medical Group	Dental surgery, eye care, cardiac interventions, orthopaedic surgery, cosmetic surgery
The InciDental Tourist	bridgework, partial and full crowns, laser tooth bleaching, implants, inlays, onlays, veneers, root canals, extractions, and other dental procedures
Tooth Tourism	Dental bonding and contouring, dental bridge, dental crowns, dental fillings, dental implants, dental veneers, dentures, gingivectomy, root canal, teeth whitening
Unbelievable India	Plastic surgery, correction of congenital malformations, reconstructive surgery, orthopedics, ophthalmology, hip replacements, Ayurvedic medicine, dental procedures, other treatments
Victus Global Healthcare	Heart bypass, heart valve replacement, hip replacement, hysterectomy, knee replacement, spinal fusion, additional treatments

Table 4. Summary of Core Marketing Messages

Company	Core Marketing Message
Axiom Health Solutions	access to affordable, timely, and high-quality care
Canadian Healthcare International (CHI)	access to affordable and high quality health care in Canada
CubaMedicare	access to affordable, timely, and high quality health care as well as vacation experience
EcuMedical Resources International Ltd.	access to timely and high-quality health care
First Choice Medical Tourism	access to affordable, timely, and high quality health care as well as holiday experience
Health Trips	access to affordable and high-quality care as well as vacation experience
Health Vacations, Inc.	access to affordable, timely, and high-quality care
International Medical Network	access to affordable, timely and high-quality care
JD Healthcare	access to affordable, timely and high-quality health care
LAM International (Logistic.Assistance.Medical International)	access to timely organ transplants
MedAsia	access to affordable, timely, and high-quality care
MedExpress Tourism	access to affordable and high-quality care
Medi-Pro Medical Management	access to timely and high-quality care
MedSolution	access to affordable, timely and high-quality care
Medtourlink	access to high-quality care
Reach Health Services & Outsourcing	Access to affordable and timely health care
Recover Discover Healthcare	access to affordable and high- quality care
Royal Med Services	access to high-quality health care
Speedy Surgery Global Healthcare	access to affordable, timely and high-quality health care
Star Hospitals	access to affordable, timely, and high-quality health care
Sun Medical Group	access to affordable and timely care
The InciDental Tourist	access to affordable and high-quality care
Tooth Tourism	access to affordable and high-quality dental care
Unbelievable India	access to affordable, timely, and high-quality care
Victus Global Healthcare	access to affordable and timely health care (messages targeted at employers rather than individuals)

reservations. Seventeen businesses advertised tours, side trips, and other holiday excursions in addition to medical care. This latter finding will not resolve disagreements about whether "medical tourism" is a suitable term for academic analysis. However, it does reveal that many medical tourism companies in Canada market both medical care and leisure or "holiday" activities commonly associated with tourism. Table 5 identifies whether medical tourism companies booked travel arrangements, arranged hotel accommodations, and offered holiday excursions, side trips, visits to local attractions, and other "tourist-like" excursions.

Table 5. Medical Tourism Companies Marketing Travel, Accommodations, and Tourism-related Activities (Y = Yes, N = No, NA = Not Addressed on company website)

Company	Book Travel	Book Hotel	Book Tours
Axiom Health Solutions	Y	Y	Y
Canadian Healthcare International (CHI)	Y	Y	NA
CubaMedicare	Y	Y	Y
EcuMedical Resources International Ltd.	Y	Y	NA
First Choice Medical Tourism	NA	NA	Y
Health Trips	Y	Y	Y
Health Vacations, Inc.	Y	Y	Y
International Medical Network	Y	Y	Y
JD Healthcare	N	NA	Y
LAM International (Logistic Assistance Medical International)	NA	Y	Y
MedAsia	Y	Y	Y
MedExpress Tourism	Y	Y	N
Medi-Pro Medical Management	Y	NA	NA
MedSolution	Y	Y	Y
Medtourlink	NA	Y	NA
Reach Health Services & Outsourcing	NA	Y	NA
Recover Discover Healthcare	NA	NA	Y
Royal Med Services	Y	Y	Y
Speedy Surgery Global Healthcare	Y	Y	NA
Star Hospitals	Y	Y	Y
Sun Medical Group	NA	NA	Y
The InciDental Tourist	N	Y	Y
Tooth Tourism	N	Y	Y
Unbelievable India (partnership with Focuz Group)	Y	Y	Y
Victus Global Healthcare	Y	Y	NA

DISCUSSION

Principal Results

Value in studying functioning and inoperative medical tourism companies

To date, there are no published studies that identify by company name, analyze in systematic fashion, disclose in transparent manner, and permanently archive websites of medical tourism companies with head offices or affiliate offices in Canada. Previous articles mention several medical travel facilitators based in Canada but do not provide a comprehensive overview of identifiable medical tourism companies, disclose their websites, analyze their business operations, and clearly distinguish between functioning and defunct companies [47]. This article provides insight into those medical tourism companies in Canada that came into existence and at some point ceased operations between 2006-2011. To some individuals, identifying and analyzing medical tourism companies that have exited the marketplace might seem like an unproductive exercise. However, identifying and analyzing medical tourism companies that have ceased functioning serves two purposes. First, it contributes to the overall analysis of Canada's medical tourism industry rather than presenting just a limited account of businesses successfully involved in advertising health care at international destinations. Analyzing companies that ceased functioning permits insight into where they were based, what health care procedures they marketed, where they proposed sending clients, what part of the marketplace they sought to occupy, and whether they advertised tourism-related activities such as offering side trips and holiday excursions in addition to booking flights and accommodations. Second, and perhaps more importantly, identifying and examining inoperative medical tourism companies challenges the hyperbolic marketing rhetoric surrounding the topic of medical tourism and could play a role in promoting more balanced ethical, social, and economic analysis of the subject [48]. Not all medical tourism companies remain going concerns. Studying websites of medical tourism companies provides no insight into how many Canadians seek care beyond Canada's borders. Nonetheless, perhaps the disappearance of an estimated 50% of all medical tourism companies based in Canada should prompt questions about the overall size of the marketplace within Canada for international medical travel. It is conceivable that the market for medical care abroad is smaller than anticipated and some medical tourism companies fail because their prospective clientele is less numerous than assumed [49]. Another possibility is that of the many medical tourism companies established in Canada, only a limited number of them have successfully developed significant client bases. It is also conceivable that some medical tourism companies were commercially successful but closed for non-financial reasons. This study does not attempt to address these questions but it should prompt critical reflection on the size and evolving shape of Canada's marketplace for medical tourism. Future research will attempt to address the question of why some medical tourism companies exit the marketplace whereas other businesses appear to be successful at attracting clients and expanding their operations.

Key findings from analysis of websites of Canadian medical tourism companies are that twenty-five companies established for the purpose of sending residents of Canada to medical facilities in other nations are no longer operational, most defunct medical tourism companies are "generalist" rather than "specialist" businesses, and many companies offer tourism-related activities in addition to marketing medical procedures.

Medical tourism company "hotspots"

Of the twenty-five companies with websites, thirteen businesses based in Ontario are now defunct, seven companies in British Columbia ceased operations, four companies in Quebec are closed, and one company in Alberta did not respond to calls and emails and is presumed closed. This study does not address how many individuals travel from Canadian provinces to international medical destinations. However, it is worth noting the concentration of defunct medical travel companies in Ontario, British Columbia, and Québec. These provinces also happen to be the three provinces with the greatest number of medical tourism companies still functioning. Perhaps they should be characterized as "hotspots" for both the establishment and closure of Canadian medical tourism companies. Future research will explore why numerous medical tourism companies have been established in some Canadian provinces whereas other provinces have no businesses involved in the medical tourism industry.

Choice of marketing messages

There likely are numerous reasons why medical tourism companies based in Canada use particular marketing messages. Most clients of these businesses must pay out-of-pocket for private medical care provided by hospitals and clinics based outside Canada. Though there are some circumstances in which provincial health insurance plans will fund provision of elective medical procedures performed outside Canada, in most instances clients of these businesses must use personal savings to purchase treatment. It is therefore unsurprisingly that affordability of care is emphasized in marketing messages. Access to high-quality care is presumably mentioned as a way to reassure prospective clients that they should feel confident that they will receive professional care when they go abroad for treatment. This message is perhaps promoted as a way of allaying fears that going abroad for inexpensive care might increase risk of exposure to lower-quality care. Finally, the message of gaining access to "timely" health care might have particular resonance for Canadians wait-listed for orthopaedic procedures, ophthalmologic procedures, spinal surgery, bariatric surgery, and other interventions where there are sometimes considerable delays prior to obtaining access to publicly-funded health care at Canadian medical facilities. For example, Speedy Surgery Global Healthcare, one defunct company mentioned in this analysis, and Timely Medical Alternatives, a company that continues sending Canadians to hospitals and clinics in the United States as well as private health care facilities in Canada, both emphasize in their company names provision of prompt access to care. Wanting to avoid providing a reductionist, single-factor account of how medical tourism is marketed by companies based in Canada, it nonetheless seems reasonable to suggest that timely access to care abroad, in contrast to treatment delays for access to some types of care in medical procedures performed in Canada, is one marketing message likely to resonate with Canadians frustrated with wait-times for various elective procedures. In this respect, it is possible that the marketing messages used by medical tourism companies based in Canada might differ somewhat from marketing claims used by medical tourism companies located in countries where the presence of a significant private health sector means that clients with sufficient financial assets can gain prompt access to medical procedures. Comparative studies that contrast marketing messages of medical tourism companies based in different countries should provide insight into regional and national variations in the types of marketing messages used to promote medical tourism.

Generalist and specialist companies

Another finding that merits mention is that twenty-two of the twenty-five defunct medical tourism companies can be characterized as "generalist" medical tourism companies. This label means that these companies marketed multiple procedures, tests, and treatments rather than confining themselves to dental treatments or some other specialized domain of care. Even companies that offer many different services are at risk of ceasing operations. It is not known whether cessation of operations is linked to a limited client base, an overreliance on one or two health care destinations, concerns about liability, competition from other medical tourism companies, a low ratio of earnings to expenditures, or other factors. Future research will explore why some companies remain operational and other businesses cease functioning.

Closure of dental tourism companies

Both of the companies that promoted "dental tourism" are no longer operational. While this article does not address why medical tourism companies ceased operations, numerous hypotheses are worth exploring in future studies. Cost savings associated with obtaining dental care abroad might be lower than anticipated once airline tickets, ground transport, hotel accommodations, and other expenses are factored into consideration. Concerns about patient safety and quality of care might deter some Canadians from receiving dental care abroad. It is also conceivable that the practice of dentistry is not well suited to a business model that involves travel to international dental clinics. Many forms of dental care require repeat visits rather than one-time procedures. Having multiple dental procedures at one time could put patients at increased risk of complications. Staying abroad for multiple procedures would presumably increase time away from home and also generate additional expenses. Though this study offers no definitive insights into why both of Canada's dental tourism companies are defunct, it is worth noting that at present there appear to be no functioning companies in Canada that are based exclusively upon a business model of marketing dental care abroad.

Little marketing of organ transplants performed abroad

Travel abroad for the purpose of purchasing a kidney used in organ transplantation, sometimes described as "transplant tourism", receives considerable attention from both health researchers and journalists. However, of Canadian medical tourism companies no longer in operation, just one business had as its primary message the marketing of organ transplants abroad. One other company marketed organ transplants in addition to other medical procedures. It is conceivable that other medical tourism companies did not market organ transplants abroad because of fears that they might be accused of facilitating "organ brokering" or "organ trafficking"; concerns about negative news media coverage that might harm other aspects of their business; inability to identify facilities able to provide organ transplants; ethical reservations about intentionally or unintentionally participating in organ trafficking; or concerns about legal repercussions. Whatever the reason, with the exception of one business that advertised organ transplants in Colombia and one company that claimed to be able to arrange transplants in India, most now-defunct medical tourism companies in Canada did not, while they remained operational, openly market organ transplants performed in other countries.

Medical consumerism and restrictions on choice of destination nations

Another noteworthy finding is that of the twenty-five businesses with websites, fourteen of them sent their clients to just one medical facility or destination nation. In this business model, owners of medical tourism companies decide that all their clients will travel to just one location. It is conceivable that this arrangement is not an optimal business model for an industry that in many respects is built upon promoting patient choice and medical consumerism. It is possible that medical tourism companies offer more attractive options to a larger prospective clientele when they market health care facilities at multiple international destinations. Whether or not this claim is credible, it is important to note that even businesses with multiple destinations ceased operations. One dental tourism company, for example, listed seven national destinations on its website. In short, while limiting the number of medical destinations constrains choice and possibly acts as a deterrent to prospective clients, promoting choice in the form of multiple travel destinations does not ensure long-term success in the medical tourism marketplace.

"Tourism" component of medical tourism

Though some researchers express reservations about the phrase "medical tourism", many now-inoperative medical tourism companies in Canada marketed travel-related services such as booking air and ground transport, organizing hotel accommodations and arranging side trips, and other tourist-like activities. In this aspect of their business models, medical tourism companies resemble travel agencies. These activities provide some insight into how medical care can be connected to activities more typically associated with "traditional" forms of tourism.

Limited media coverage of defunct medical tourism companies

Whatever causes medical tourism companies to close, it is important to note that in Canada a substantial number of medical tourism companies have exited the marketplace. The closure of approximately half of all medical tourism companies in Canada has not been previously examined and addressed by health researchers and journalists. Whereas medical tourism companies routinely issue press releases and attract news media coverage when they commence operations, these businesses typically do not issue press releases or receive news coverage when they close. Just one business, EcuMedical Resources

International, generated news coverage as it ceased functioning [50, 51]. In that company's case, several of its clients claimed that they paid EcuMedical for arranging medical procedures in the U.S. and then began receiving bills from the hospitals where they received care. Following a police investigation the two company owners were charged with multiple counts of fraud. In this case, activity that is presumably atypical in the medical tourism industry resulted in considerable local news coverage of a medical tourism company ceasing operations.

Basis for additional research

One important benefit of studying medical tourism companies by analyzing content of medical tourism company websites is that the exercise can challenge hitherto unexamined assumptions about the industry. More fine-grained analyses that draw upon multiple research methods such as surveys, in-depth interviews, analysis of news media coverage, and participant observation should generate additional insights into medical tourism companies in Canada. For example, quantitative studies should provide insight into how many Canadians go abroad for care, where they seek treatment, and what types of procedures they select. Identifying Canada's medical tourism companies, providing their websites, describing their operations and business models, charting where they send patients abroad, summarizing the procedures they advertise, and explicating what services they offer in addition to organizing health care packages should facilitate further studies into Canada's medical tourism industry. It should also be of practical use to scholars interested in comparing medical tourism companies in Canada with similar businesses situated in other countries. For example, researchers in Australia, England, the United States and elsewhere might find it helpful to compare and contrast Canada's medical tourism industry with the emergence of medical tourism facilitators in their home countries. In addition, these findings should assist scholars interested in evaluating the quality of health-related information presented on the websites of medical tourism companies, and the use of YouTube, Twitter, Facebook, and various forms of social media by medical tourism companies situated in Canada. Scholars interested in using various research methods to study medical tourism companies in Canada should now have a better sense of where to focus their attention. To facilitate research by other scholars, websites or printouts of websites analyzed in this article have been electronically archived in a publicly accessible database.

Limitations

Findings limited to medical tourism companies based in Canada

This article is based upon content analysis of websites of medical tourism companies with home offices or affiliate offices in Canada. It does not provide insight into medical tourism companies located in other nations. This study offers a set of categories or framework that can be used to study medical tourism companies in other nations. However, using content analysis to study websites of medical tourism companies based outside Canada might generate results very different from this study's findings.

Possibility of undiscovered websites

Internet searches, review of news reports, review of government records of federally incorporated companies in Canada, and examination of existing lists of medical tourism companies in Canada were all used in an attempt both to identify the total number of medical tourism companies established in Canada and distinguish medical tourism companies that are currently functioning from companies that are no longer involved in sending clients abroad for medical care. However, it is possible that some medical tourism companies were not identified in these searches. Medical tourism companies use company websites, Twitter, YouTube, Facebook, and other media to market international care, and internet searches are therefore a useful tool to use when trying to find medical tourism companies. Nonetheless, it is conceivable that some medical tourism companies based in Canada do not advertise using the internet or that repeated searches overlooked some businesses.

Possibility that non-responders might remain in business

Companies that did not respond to seven phone calls and/or emails over a three week period of repeated queries were listed as defunct. Consultants to the medical tourism industry recommend that calls to medical tourism companies should be promptly returned within twenty-four hours and preferably within two hours [52]. However, it is possible that some medical tourism companies continue to function even though they did not reply to multiple efforts to contact them. Establishing that businesses have ceased functioning in the absence of obvious signs such as companies losing their status as federally incorporated businesses in the end involved exercising judgement.

Investigative research methods might challenge claims made by medical tourism company websites

The study is limited insofar as it analyzes content found on medical tourism company websites and does not question the veracity of various claims. For example, one medical tourism company website states that clients can select clinics from among seven different countries. Someone using the tools of investigative journalism might discover that companies do not have business relations with all the facilities that they mention as international health care destinations. Such questions were bracketed from this study but they are worth pursuing in other studies. It is conceivable that some companies make exaggerated claims about destination facilities and advertised medical services and do not have experience coordinating particular health care packages and sending clients to advertised destinations. Content analysis, for the purpose of this article, accepted statements made by medical tourism companies. Such an approach, though it provides insight into how medical tourism companies in Canada advertise their services, would benefit from being complemented by investigative approaches that critically examine the credibility of claims made by medical tourism companies.

Studying Defunct Companies

This article focuses upon medical tourism companies that ceased operations. Performing a content analysis of websites of defunct companies could be seen as a limitation of the study, though the point of the exercise was to investigate and analyze not only medical tourism companies that are flourishing in a highly competitive marketplace but also to analyze websites of companies that have ceased sending clients to international health care destinations.

Comparison With Prior Work

Focus on Canada

To date, most studies of medical tourism companies have focused upon websites of medical tourism facilitators based in the United States. This paper makes a meaningful contribution to scholarship by providing content analysis of medical tourism companies with home offices or affiliate branches in Canada. The local, regional, and national settings within which medical tourism companies are situated matters. It should not be assumed that medical tourism companies in Canada advertise the same health care packages or target the same prospective client base as businesses in Australia, the United States, and elsewhere.

Disclosure of Primary Sources

This article differs from previous work on medical tourism companies by identifying particular businesses, performing content analysis of their websites, and enabling readers to compare this analysis with the primary sources of medical tourism company websites. This step, by linking content analysis to identified primary sources, promotes transparency in research and allows other scholars to corroborate or challenge findings described in this study. Identifying particular companies and providing content analysis of their websites might help other researchers use their disciplinary methods and theories to study Canada's medical tourism industry.

Analyzing Company Websites

This article can be distinguished from most prior work on medical tourism due to its emphasis on content analysis of medical tourism company websites. Medical tourism companies conduct much of their business using the internet [53, 54, 55]. Reliance on the internet as a medium for communication, marketing, and interaction with clients means that much can be learned about medical tourism companies through careful analysis of their websites. Scholarly analysis of medical tourism companies should not be confined to content analysis of medical tourism company websites. Nonetheless, this mode of research makes an important contribution to the study of transnational medical travel.

Studying Inoperative Medical Tourism Companies

Finally, to date scholarship on medical tourism companies has examined companies currently engaged in the business of sending clients to international health care destinations. This article differs from previous work by identifying medical tourism companies that have ceased functioning. Identifying medical tourism companies that are inoperative raises questions about the nature and future direction of this industry. Many commentators assume that the medical tourism industry is in a period of unprecedented growth [56]. This article suggests that while the overall medical tourism industry might be expanding, the long-term success of individual medical tourism companies is not assured. The closure of twenty-five medical tourism companies in Canada might reflect a period of industry consolidation in which some companies flourish while others disappear or are subsumed by other companies. It is also possible that interest within Canada for travelling abroad to international medical facilities is not as large as is commonly assumed and these companies ceased functioning because many businesses are competing for a limited client base. Another possibility is that many of the companies erred when they based their operational model on sending clients to just one medical facility or country. It is conceivable that prospective clients want to have increased choice in where they travel for care and a business model that constrains options serves to deter clients that might otherwise consider going abroad for health care. And finally, since individuals considering going abroad for medical care can use the Internet to gain direct access to international hospitals and clinics, it is possible that medical tourism companies play a relatively minor role in the overall marketplace for medical travel. There might be less of a need for the service of medical travel facilitation than many entrepreneurs assume. Future research will explore in greater detail why the medical tourism companies identified in this study ceased functioning.

CONCLUSIONS

This study identifies twenty-five defunct Canadian medical tourism companies. Researchers will likely disagree concerning how to interpret some aspects of these findings. Individuals supportive of the Canada Health Act and publicly funded access to medically necessary health services and critical of the privatized, for-profit health care marketed by medical tourism companies might argue that the failure of twenty-five medical tourism companies is a sign of limited interest within Canada for medical travel. They might claim that news media coverage of the phenomenon of medical tourism far outpaces the actual significance of the topic. In contrast, proponents of for-profit, private health care might assert that in every competitive marketplace some companies prosper while other businesses disappear. Whatever lessons can be extracted from the closure of twenty-five medical tourism companies established in Canada, it seems reasonable to suggest that establishing a medical tourism company in Canada provides no assurance of commercial success.

This study accomplishes several goals. Medical tourism companies use the internet to market health services to prospective clients. Their websites provide valuable information about where medical tourism companies are located within Canada, what countries and facilities they select as destination sites, what health services they advertise, how they position themselves in a competitive national and global marketplace, and what travel-related services they promote in addition to marketing health care. Content analysis of medical tourism company websites permits development of a detailed account of the current state of Canada's medical tourism industry. This article reviews defunct medical tourism facilitators in Canada. Websites of operational medical travel companies based in Canada are examined elsewhere. Various research methods and theories can be used to build upon the analysis provided here.

Many researchers lament the lack of empirical research into the study of medical tourism. This article addresses these concerns and opens new avenues for the study of medical tourism companies, medical travelers, and international health care destinations. It also reveals the importance of studying medical tourism in Canada and elsewhere, and not making the error of thinking that the medical tourism industry is primarily a product of the large population of uninsured individuals in the United States. Many researchers and other interested parties will likely be surprised to learn just how many medical tourism companies have emerged in Canada, a country where residents have publicly funded access to a large basket of medically necessary health services. They might be equally surprised to learn how many of these businesses have come and gone and no longer market transnational medical travel to Canadians. Establishing just how many residents of Canada use the services promoted by medical tourism companies based in Canada, examining what happens to medical travelers when they go abroad for care, and learning more about their experiences once they return to Canada will have to await additional studies by researchers from a variety of disciplines. This study attempts to make a substantial contribution to the study of the Canadian medical tourism industry. It also provides sign posts that should be of practical use to health researchers and other individuals interested in exploring ethical, legal, social, cultural, economic, and public health implications of Canada's medical tourism industry.

AUTHOR'S CONTRIBUTIONS

LT is the sole author of this article. He designed the study, conducted content analysis of medical travel company websites, and wrote the manuscript.

AUTHOR'S INFORMATION

The author studies ethical and social issues related to transnational medical travel and the emergence of a global marketplace in health services.

ACKNOWLEDGEMENTS AND FUNDING

Study of Canada's medical tourism industry was facilitated by a Distinguished Visiting Fellowship (2006-2007) provided by the Comparative Program on Health and Society, Munk School of Global Affairs, University of Toronto. The Lupina Foundation provides funding for the Comparative Program on Health and Society. Analysis of Canadian medical tourism companies was also enabled by a William Dawson Scholar award (2006-2008) and additional sources of funding provided by McGill University. I wish to express appreciation for these sources of financial support. In addition, I wish to thank former colleagues at the University of Toronto as well as within the Biomedical Ethics Unit at McGill.

COMPETING INTERESTS

The author declares that he has no competing interests.

REFERENCES

1. Connell J: Medical tourism: Sea, sun, sand and...surgery. Tourism Management. 2006, 27: 1093-1100.

2. Garcia-Altes A: The development of health tourism services. Annals of Tourism Research. 2005, 32 (1): 262-266.

3. Goodrich JN, Goodrich GE: Health care tourism-an exploratory study. Tourism Management. 1987, 217-222.

4. Healy C: Surgical tourism and the globalization of healthcare. Ir J Med Sci. 2009, 178: 125-127.

5. Crooks V, Snyder J: Medical tourism: What Canadian family physicians need to know. Canadian Family Physician. 2011, 57: 527-529.

6. MacReady N: Developing countries court medical tourists. Lancet. 2007, 369: 1849-1850.

7. Ramirez de Arellano AB: Patients without borders: The emergence of medical tourism. IJHS. 2007, 37 (1): 193-198.

8. Unti JA: Medical and surgical tourism: The new world of health care globalization and what it means for the practicing surgeon. Bulletin of the American College of Surgeons. 2009, 94 (4): 18-25.

9. Behrmann J, Smith E: Top 7 issues in medical tourism: challenges, knowledge gaps, and future directions for research and policy development. Global Journal of Health Science. 2010, 2 (2): 80-90.

10. Lunt N, Carrera P: Medical tourism: assessing the evidence on treatment abroad. Maturitas. 2010, 66: 27-32.

11. Priest L: Sold! Triple bypass surgery in U.S. for $16,000. The Globe and Mail. 2011

12. Hopkins L, Labonte R, Runnels V, Packer C: Medical tourism today: what is the state of existing knowledge?. Journal of Public Health Policy. 2010, 31: 185-198.

13. Crooks VA, Kingsbury P, Snyder J, Johnston R: What is known about the patient's experience of medical tourism? A scoping review. BMC Health Services Research. 2010,

14. Carrera P, Bridges JF: Globalization and healthcare: understanding health and medical tourism. Expert Rev. Pharmacoeconomics Outcomes Res. 2006, 6 (4): 447-454.

15. Kaspar C: A new lease on life for spa and health tourism. Annals of Tourism Research. 1990, 17 (2): 298-299.

16. Turner L: 'Medical tourism' and the global marketplace in health services: U.S. patients, international hospitals, and the search for affordable health care. IJHS. 2010, 40 (3): 443-467.

17. Milstein A, Smith M: America's new refugees-seeking affordable surgery offshore. NEJM. 2006, 355 (16): 1637-1640.

18. Schiano TD, Rhodes R: The dilemma and reality of transplant tourism: an ethical perspective for liver transplant programs. Liver Transplantation. 2010, 16: 113-117.

19. Kiatpongsan S, Sipp D: Monitoring and regulating offshore stem cell clinics. Science. 2009, 323: 1564-1565.

20. Smith E, Behrmann J, Martin C, Williams-Jones B: Reproductive tourism in Argentina: clinic accreditation and its implications for consumers, health professionals and policy makers. Developing World Bioethics. 10 (2): 59-69.

21. Huxtable R: The Suicide tourist trap: compromise across boundaries. Journal of Bioethical Inquiry. 2009, 6 (3): 327-336.

22. Turner L: 'Medical tourism' initiatives should exclude commercial organ transplantation. J R Soc Med. 2008, 101 (8): 391-394.

23. Burkett L: Medical tourism: concerns, benefits, and the American legal perspective. Journal of Legal Medicine. 2007, 28: 223-245.

24. Mattoo A, Rathindran R: How health insurance inhibits trade in health care. Health Affairs. 2006, 25 (2): 358-368.

25. Smith RD, Chanda R, Tangcharoensathien V: Trade in health-related services. Lancet. 2009, 373: 593-601.

26. Crozier GK, Baylis F: The ethical physician encounters international medical travel. J Med Ethics. 2010, 36: 297-301.

27. Turner L: 'First world health care at third world prices': Globalization, bioethics and medical tourism. BioSocieties. 2007, 2: 303-325.

28. Whittaker A, Manderson L, Cartwright E: Patients without borders: understanding medical travel. Med Anthropol. 2010, 29 (4): 336-343.

29. Johnston R, Crooks VA, Snyder J, Kingsbury P: What is known about the effects of medical tourism in destination and departure countries? A scoping review. International Journal for Equity in Health. 2010, 9: 24-

30. Alleman BW, Luger T, Reisinger HS, Martin R, Horowitz MD, Cram P: Medical tourism services available to residents of the United States. J Gen Intern Med. 2011, 26 (5): 492-497.

31. Shah KY: Current status and emerging trends of international medical outsourcing in the United States--a qualitative study. 2007, University of South Carolina doctoral dissertation, [http://www.webcitation.org/5x9wiGYUu]

32. Sobo EJ, Herlihy E, Bicker M: Selling medical travel to US patient-consumers: the cultural appeal of website marketing messages. Anthropology & Medicine. 2011, 18 (1): 119-136.

33. Johnston R, Crooks VA, Adams K, Snyder J, Kingsbury P: An industry perspective on Canadian patients' involvement in Medical Tourism: implications for public health. BMC Public Health. 2011, 11: 416-

34. Turner L: Medical tourism: Family medicine and international health-related travel. CFP. 2007, 53: 1639-1641.

35. Snyder J, Crooks V, Adams K, Kingsbury P, Johnston R: The patient's physician one-step removed': the evolving roles of medical tourism facilitators. Journal of Medical ethics. 2011, 37: 530-534.

36. Blackwell T: Cuba draws Canadian patients: Orthopedic, visual operations bring foreigners. National Post. 2007, A9-

37. McMurdy D: Vanity takes a vacation. National Post. 2007, FW.2.

38. Raman N: Medical travel agency lifts off. Montreal Gazette. 2006

39. McGinn D: Sun, sand and surgery. Globe and Mail. 2008

40. Ward D: Long wait times drive Canadians overseas for medical treatment. Edmonton Journal. 2007

41. Floyd M, Izenberg D, Kelly B, et al: Medical services directory. Maclean's. 2006

42. Find Medical Travel Services. [http://www.findinghealthcare.ca/medical_tourism.html]

43. Medical Tourism Facilitators. Tourism Review.com. 2009, [http://www.tourism-review.com/fm956/m6.pdf]

44. Pope C, Ziebland S, Mays N: Qualitative research in health care: Analysing qualitative data. BMJ. 2000, 320: 114-116.

45. Industry Canada. Corporations Canada, [https://www.ic.gc.ca/app/scr/cc/CorporationsCanada/fdrlCrpSrch.html;jsessionid=0000v9O3VPcCrmqiXhaZKPaOE2o:15dmaulj8?locale=en_CA]

46. Tri-Council Policy Statement 2. [http://www.pre.ethics.gc.ca/pdf/eng/tcps2/TCPS_2_FINAL_Web.pdf]

47. Turner L: Canadian Medicare and the Global Health Care Bazaar. Policy Options. 2007

48. Keckley PH, Underwood HR: Medical Tourism: Consumers in Search of Value. 2008, Deloitte Center for Health Solutions, [http://www.deloitte.com/assets/Dcom-UnitedStates/Local%20Assets/Documents/us_chs_MedicalTourismStudy(3).pdf]

49. Johnson TJ, Garman AN: Impact of medical travel on imports and exports of medical services. Health Policy. 2010, 98 (2-3): 171-177.

50. Puzic S: Cancer patient bilked in alleged fraud. Windsor Star. 2010

51. Puzic S: Windsor police lay fraud charges against EcuMedical couple. Windsor Star. 2010

52. Passmore R: Converting international patient inquiries @10%-Part 2. Medical Tourism Magazine. 2009

53. Cormany D, Baloglu S: Medical travel facilitator websites: An exploratory study of web page contents and services offered to the prospective medical tourist. Tourism Management. 2011, 32: 709-716.

54. Lunt N, Hardey M, Mannion R: *Nip, tuck and click*: Medical tourism and the emergence of web-based health information. The Open Medical Informatics Journal. 2010, 4: 1-11.

55. Mason A, Wright KB: Framing medical tourism: An examination of appeal, risk, convalescence, accreditation, and interactivity in medical tourism web sites. Journal of Health Communication. 2011, 16 (2): 163-177.

56. York D: Medical tourism: The trend toward outsourcing medical procedures to foreign countries. Journal of Continuing Education in the Health Professions. 2008, 28 (2): 99-102.

3
CHAPTER

Medical Tourism: A Cost or Benefit to the NHS?

Johanna Hanefeld[1*], Daniel Horsfall[2], Neil Lunt[2], Richard Smith[3]

Department Global Health and Development, Faculty of Public Health and Policy, London School of Hygiene and Tropical Medicine, London, United Kingdom

Department of Social Work and Social Policy, University of York, York, United Kingdom

Department of Social Work and Social Policy, University of York, York, United Kingdom

Faculty of Public Health & Policy, London School of Hygiene & Tropical Medicine, London, United Kingdom

ABSTRACT

'Medical Tourism' – the phenomenon of people travelling abroad to access medical treatment - has received increasing attention in academic and popular media. This paper reports findings from a study examining effect of inbound and outbound medical tourism on the UK NHS, by estimating volume of medical tourism and associated costs and benefits. A mixed methods study it includes analysis of the UK International Passenger Survey (IPS); interviews with 77 returning UK medical tourists, 63 policymakers, NHS managers and medical tourism industry actors policymakers, and a review of published literature. These informed costing of three types of treatments for which patients commonly travel abroad: fertility treatment, cosmetic and bariatric surgery. Costing of inbound tourism relied on data obtained through 28 Freedom-of-Information requests to NHS Foundation Trusts. Findings demonstrate that contrary to some popular media reports, far from being a net importer of patients, the UK is now a clear net exporter of medical travellers. In 2010, an estimated 63,000 UK residents travelled for treatment, while around 52,000 patients sought treatment in the UK. Inbound medical tourists treated as private patients within NHS facilities may be especially profitable when compared to UK private patients, yielding close to a quarter of revenue from only 7% of volume in the data examined. Costs arise where patients travel abroad and return with complications. Analysis also indicates possible savings especially in future health care and social costs averted. These are likely to be specific to procedures and conditions treated. UK medical tourism is a growing phenomenon that presents risks and opportunities to the NHS. To fully understand its implications and guide policy on issues such as NHS global activities and patient safety will require investment in further research and monitoring. Results point to likely impact of medical tourism in other universal public health systems.

INTRODUCTION

The phenomenon of people travelling abroad to access medical treatment – commonly termed 'Medical Tourism' – has received increasing attention in academic and popular media [1]. The confluence of available and affordable air travel, internet-based marketing by providers, and an increasing requirement for out-of-pocket expenditure, even in universal public health care systems such as the UK NHS, suggests that increasing numbers of patients may consider travelling for treatment. The PIP scandal highlighted challenges for UK patients in seeking redress from private providers, especially where these may be based in other jurisdictions [2],[3].

As the new NHS reforms introduce yet greater market elements, including the removal of the cap on income from private patients [4], and the EU Directive on crossborder healthcare is implemented which codifies rights around patient mobility [5], it is imperative to consider the challenges and opportunities that medical tourism – inward and outward – may present to the NHS [6].

Yet, reliable information on even the basic number, characteristics, motivations and experiences of such patients is scarce, as patients arrange and pay for such care privately [7]. Indeed, a recent review of medical tourism literature [8] found that academic literature relies heavily on opaque data from private consultancy firms or unverified media reports [9], [10]. In the absence of even the basic level of information in these areas, it is understandable that rhetoric has filled the vacuum. In this paper we present evidence from the largest study yet conducted concerning medical tourism, undertaken from an NHS perspective, to provide a firmer footing for debate and discussion by health professionals, NHS managers and those involved in the wider policy-making context.

METHODS

Authors interviewed 77 UK medical tourists and 63 other UK stakeholders between March 2011 and August 2012. Interviewees gave written consent to participate in the study. Interviews were recorded, transcribed and thematically analysed. The study received ethical clearance from the National NHS Ethics review process submitted through the Sheffield Research Ethics Committee approval (11/H1308/3).

Analysis is three-fold: (i) the volume and characteristics of outbound and inbound UK medical tourists is based upon the International Passenger Survey (IPS); (ii) assessment of NHS income from foreign patients is based upon freedom-of-information requests submitted to 28 NHS Foundation Trust hospitals; and (iii) evaluation of the challenges encountered, costs incurred and potential savings for the NHS is based on a review of published and grey literature and interviews with UK nationals, NHS managers and policy makers. Each of these is described below.

Analysis of the International Passenger Survey (IPS)

The IPS, conducted by the UK Office of National Statistics (ONS), collects information from passengers as they enter or leave the UK. Passengers are randomly selected as they travel through passport control and a brief survey is administered. One of the survey questions asks passengers to define their primary purpose for travel; 'medical treatment' is one of the answers recorded, thus providing insight into the number of passengers who self-declare that they are travelling for medical treatment.

The IPS dataset from 2000–2010, from the Office of National Statistics (ONS), was analysed by two authors independently, triangulating results. Data from the IPS, interviews, literature and NHS tariffs were used to calculate cost impacts. Authors used the different data sources accessed to carefully triangulate and better understand the reliability of the data from the IPS, which is reflected on in the discussion.

FOI Requests

Submitted to 28 Foundation Trust hospitals on volume and income from international private patients. Trusts were purposely selected to be those most likely to be visited by inbound tourists i.e., large and well-known Trusts, such as Great Ormond Street Hospital for Sick Children, many of which are based in London. Data on foreign patients was analysed to understand the potential of earnings from foreign patients.

Qualitative Analysis

Authors interviewed 77 UK residents who travelled abroad for treatment and 63 other UK stakeholders between March 2011 and August 2012. Interviews were recorded, transcribed and thematically analysed. The study received ethical clearance from the National NHS Ethics review process.

RESULTS

While the level of inward travel of foreign patients to the UK (although not necessarily the NHS) has been relatively stable over the last decade, there has been a substantial increase in the number of UK residents travelling abroad to access medical treatment, as indicated in Figure 1.

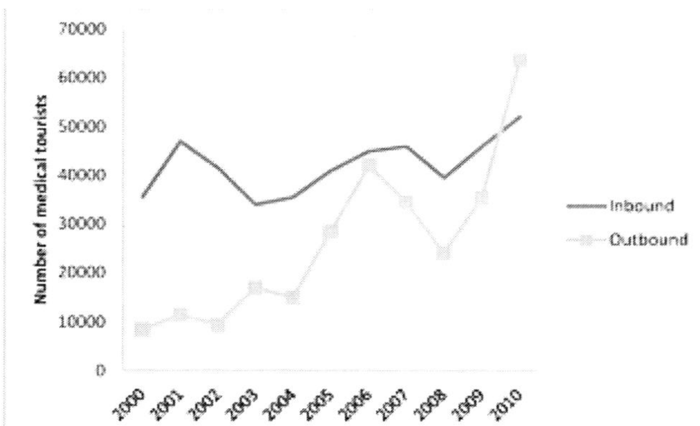

Figure 1. The number of people who travelled into or out of the UK for medical treatment during the period 2000–2010.

Destination of UK Outbound Medical Travellers

Figure 2 shows UK residents most commonly travel for medical treatment to North, West, and Southern Europe with France being the most visited country over the decade.

Figure 2. Map depicting total numbers of medical travellers and their destinations from the UK over the period 2000–2010.

Examining this in greater detail (Fig 3) suggests that Central and Eastern Europe are second most popular, and that Poland and Hungary are increasingly popular.

South Asia, primarily India, also attracts large numbers of UK patients, making it the most frequently visited non-European country, with a relatively stable pattern of travel to India, Pakistan, and in much lower numbers Sri Lanka and Bangladesh, possibly reflecting a diaspora effect. In contrast, East Asia shows a different pattern, with virtually no medical travellers recorded by the IPS prior to 2003, yet by 2010 15% of all UK medical travellers went to East Asia. This increase of 430% is unlikely to be solely related to diaspora patients, but does correlate with many South East Asian countries marketing strategies at this time [11].

Based on the IPS data, and patient interviews, treatment specific destinations emerge. For example, UK dental patients increasingly travel to Hungary and Poland, which corresponds to the varied availability of NHS dental treatment over the last decade [12]. Fertility tourists often travel to countries in Eastern Europe, Cyprus and Spain possibly owing to more easily accessible gametes, and less stringent regulation which allows anonymous donation as well as a greater number of embryos transferred [13].

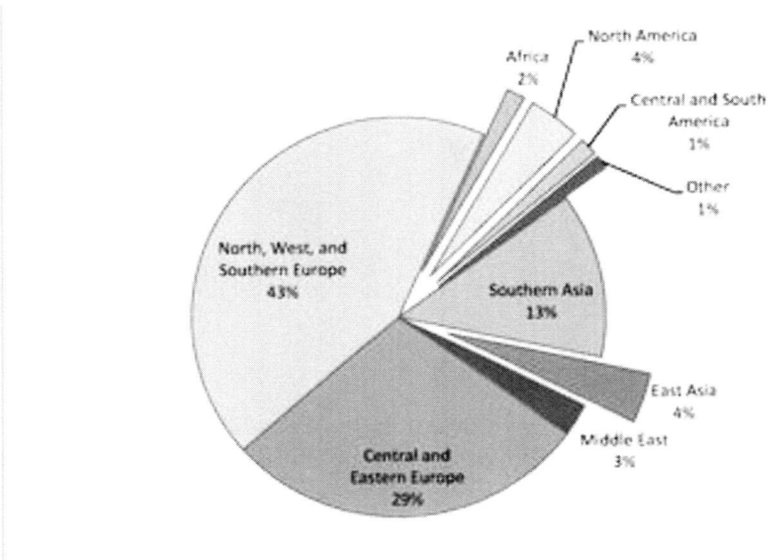

Figure 3. Pie Chart showing total outward medical travel by UK residents by destination region over the time-period 2000–2010.

Inward Medical Travel

As evident in Figure 1, data from the IPS suggests that international patient inflows to the UK (independent sector and NHS private services) were in the region of 52,000 in 2010. Data over the decade also confirms that while growing, the overall numbers of patients travelling into the UK to access medical services is rising at a much slower rate than UK residents travelling out for care. So, contrary to some popular media reports, far from being a net importer of patients, the UK is now a clear net exporter of medical travellers.

Major source countries for patients coming into the UK include Spain, Greece, Cyprus and the Middle East. The number of Greeks and Cypriots travelling into the UK to access treatment rose rapidly in 2009 and 2010. These figures may reflect a change as a result of the economic

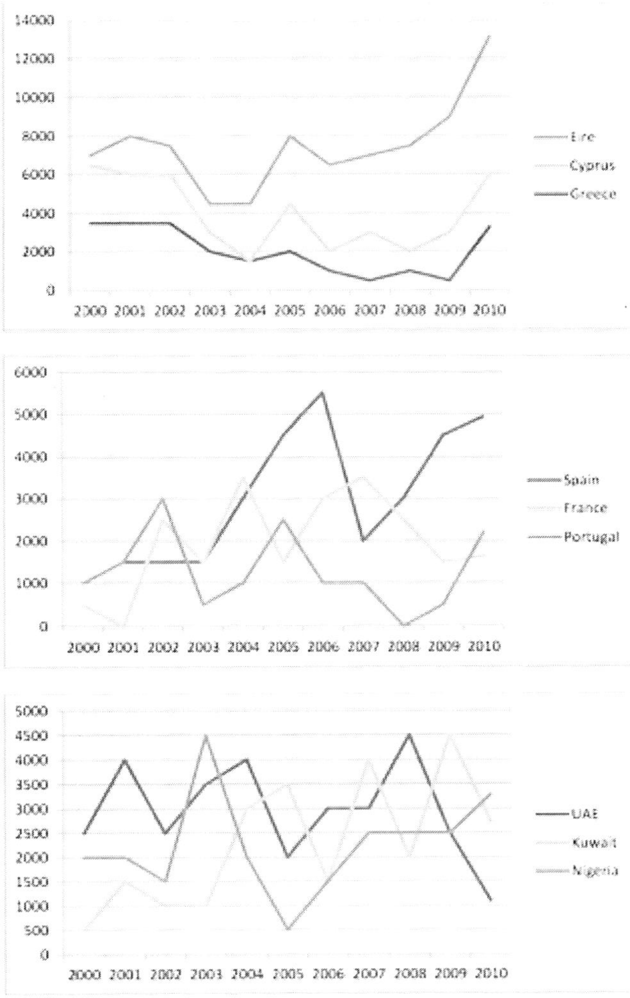

Figure 4. Nine most common countries of origin for those who travelled to the UK for medical (2000–2010).

crisis, which in turn has meant severe public sector cuts in these countries, including in health[14]. Similarly, while medical tourists from Ireland may choose to travel to access treatment not available there, including termination of pregnancies, the rapid increase in patients from Ireland in recent years may reflect the cuts in the health sector there and greater numbers of UK citizens resident in Ireland returning to the UK for treatments (see Figure 4). The 'dip' in both inbound and outbound medical travel evident in Figure 1 in 2008 may be attributable to the onset of the crisis. Examining the number of travellers by quarter found a much lower number of inward and outward medical travellers in Quarter 3 of 2008 during the onset of the crisis, than the rest of the year, or Quarter 3 in 2009. In the case of Irish, Spanish (and perhaps French) residents, it is highly likely that a substantial number will be UK expats and it is unclear whether these engage in out-of-pocket medical treatment (in the private sector or NHS) or whether they accept NHS services free at the point of use for which they may (or may not) be eligible.

A further significant number of patients travel from the Middle East (specifically from the United Arab Emirates and Kuwait) although visitor numbers from both countries dropped sharply in 2008 and 2009 respectively. Despite some variation between years, a stable inward flow of medical travellers from Nigeria is also evident over the past decade, perhaps reflecting the growing wealth of some sections of that population.

International activity within hospital trusts.

Our Freedom-of-Information requests suggest that Trusts could not always clearly identify international patients within their pool of private patients because nationality was not recorded when they underwent treatment and nationality/place of residence may differ. Looking at the 28 Trusts within our sample, their international activity ranged from relatively marginal to being one-third of their total private work.

Where Trust managers were interviewed (at seven sites) they spoke of international patient flows and activities within the context of pressure on NHS resources, and pre-existing international activities and linkages. Commercial imperatives were balanced with strong statements regarding the core NHS role, centred on NHS services and prioritising NHS patient care. International patient activity was typically specialist where it was not possible to treat locally because of relatively small volumes and the complex nature of treatment required. Relationships, primarily clinical ones, for example where a clinician from aboard had trained or worked in a UK hospital were paramount in maintaining flows of international patients. Established practices of education, training, consultancy and linkages were reported to help facilitate referrals. Rather than systematic links these personal networks appeared paramount in linking UK hospitals to international patients.

What is the Impact of Medical Tourism on the NHS 'Bottom-line'?

Using the IPS data, analysis from interviews with medical tourists, academic literature and published NHS data we calculated possible costs and savings for the NHS for three types of medical tourism identified (see annex S1 for calculations).

Fertility tourism.

Based on data from the Office of National Statistics on multiple births in the UK and evidence from a hospital in London which found over a quarter of multiple births were in women who had travelled abroad for fertility treatment [15], we estimated the cost incurred through multiple births as a result of individuals travelling abroad for fertility treatment. Multiple pregnancies pose risks to mothers and children. We concentrated on the actual costs of multiple births per se as the exact needs throughout pregnancy and possible complications are highly variable between women, and thus our estimates will be highly conservative. We calculated the additional cost of a twin or triplet over singleton birth resulting from fertility travel in 2010 to be £15.5 million.

The long-term costs resulting from assisted reproductive technologies, including multiple pregnancies will not differ between medical tourists and fertility patients who received care in the UK. However, our research indicates that patients will travel in search of reproductive care to countries with regulations that will allow fertility treatment likely to result in a higher number of multiple births. Any effort to address the rise in multiple births in the UK therefore needs to take account of medical travel and involve specific targeted information to be effective.

Cosmetic tourism.

We also calculated the likely cost of complications resulting from cosmetic tourism based on a recent study by Miyagi et al. [16], who described a cohort of patients in a tertiary facility which reported problems arising from cosmetic surgery undertaken abroad over a period of three years. The authors calculated the cost of treatment provided within the NHS for complications and highlighted the reimbursement received by the hospital from the PCT (which was less than the expenditure of the hospital). Based on our calculations complications of medical tourists are at a cost of £8.2 million per annum within the NHS.

Bariatric surgery.

Compared to other types of tourism discussed, bariatric tourism may represent savings rather than costs for the NHS, as well as wider social savings. With 25% of the UK population classified as clinically obese, the financial impact of obesity on the NHS is calculated as £4.3billion by the DoH [17]. Obesity also has wider costs for social services. For example, a study by the National Office of Accounting estimated that 18 million working days were lost due to obesity with surgery offering potential savings. Hawkins et al. [18] demonstrated that there was a 32% increase in bariatric patients in paid work after surgery.

Based on these estimates, the 13 bariatric tourists interviewed for this research would represent a saving of £112,506 (in cost of procedure and in future health care and social services savings). Even as a high estimate, the key point remains that patients travelling abroad to receive bariatric surgery are likely to represent a saving to the NHS and social services. Further research on the longitudinal effects of bariatric surgery is needed and now underway in the University of Glasgow at the Surgical Obesity Treatment Study (ScOTS).

Income Generated by Inbound Medical Travellers

Income generated by inbound medical travellers can be divided into additional tourism revenue, capturing the general expenditure related to patients visit to the UK, and medical expenditure (revenue to hospital).

Tourism revenue from all inbound medical travelers.

Tourism revenue by medical travellers to the UK per annum is based on the most recent IPS data for inbound medical travellers (2010). As respondents in the IPS survey specifically state they are visiting for health care, it is assumed they would not have otherwise have visited the UK, and thus are an addition to visitor/tourist numbers to the UK. Hence, any spending would be seen to be a net benefit not otherwise coming to the UK.

Based on hospital data for patients treated within NHS hospitals, it can be assumed that 20% of inbound medical travellers receive treatment as inpatients, the remainder as day-case procedures. Expenditure was calculated for patients staying in the UK for a number of different scenarios, ranging from those who stay for four days to receive outpatient treatment to patients who receive in-patient treatment for

10 days and stay a further two weeks for follow-up (see Table 1). Assumptions were based on interview data collected and on an average hospital stay of inpatients (not just medical tourists) in 2010 to 2011 from the NHS Hospital Episode Statistics. These assumptions were that: patients likely arrive some days before treatment and remain additional days to fully recuperate or even take the opportunity for additional tourism activities; people travel with one companion, and travellers from the Middle East travel with two, and that these are not captured by the IPS (based on interview data and corroborated by a 2008 national survey conducted by *Which?*). This seems reasonable given the higher foreign patient number captured from the FOI letters and interview data from patients who often reported reluctance to be identified as medical tourists possibly due to the negative public image of medical tourism, making it unlikely that accompanying persons will identify as medical tourists. Cost of accommodation was calculated at £80 per night and £100 per day as spending for patients when they were not in hospital and for their travel companions.

Table 1. Calculation of additional spend by incoming medical tourists and their travel companions.

Inbound medical travellers	No	Nights in hotel	Cost hotel	Expenditure	Total expenditure
	52000				
Inpatients*	10400				
Hospital for ten days 75% (75% from ME)	7800	14	8,736,000	10,920,000	19,656,000
Hospital for five days 25% (5% from ME)	2600	7	1,456,000	1,820,000	3,276,000
Subtotal inpatients			10,192,000	12,740,000	22,932,000
Accompanying persons inpatients					
Hospital for ten days 75% (75% from ME)	13650	24	26,208,000	32,760,000	58,968,000
Hospital for five days 25% (5% from ME)	2730	12	2,620,800	3,276,000	5,896,800
Subtotal accompanying persons			28,828,800	36,036,000	64,864,800
Total inpatient and accompanying			39,020,800	48,776,000	**87,796,800**
Outpatients	41600				
4 day stay (25%)	10400	4	3,328,000	4,160,000	7,488,000
7 day stay (40) (2.75% ME)	16640	7	9,318,400	11,648,000	20,966,400
14 day stay (35%) (2%ME)	14560	14	16,307,200	20,384,000	36,691,200
Subtotal outpatients			28,953,600	36,192,000	**65,145,600**
Accompanying persons outpatients	[illegible]				
4 day stay (25%)	10400	4	3,328,000	4,160,000	7,488,000
7 day stay (40) (2.75% ME)	17098	7	9,574,880	11,968,600	21,543,480
14 day stay (35%) (2%ME)	[illegible]	14	16,586,080	20,732,600	37,318,680
Subtotal accompanying persons OP	[illegible]		[illegible]	[illegible]	**66,150,160**
Total outpatient and accompanying					**131,495,760**
Total					**219,291,560**

doi:10.1371/journal.pone.0070406.t001

Calculations are summarized in Table 1 and further explained in Annex S1 suggest that, even without taking the cost of the actual medical treatment into account, medical tourists to the UK contribute around £219 million in additional 'tourism spending' to the UK economy per year.

Healthcare revenue from all inbound medical tourists.

To estimate the spend on medical procedure by inbound medical tourists in NHS facilities as accurately as possible, we submitted Freedom-of-Information requests for data on income from private patients in NHS hospitals, including UK and non-UK patients, to 28 NHS Foundation hospitals. Of 28 hospitals 19 were able to provide data on the percentage of income that resulted from non-UK resident patients and number of non UK residents treated as private patients. Authors excluded Moorfields Eye Hospital, as a review of the data across different hospitals indicated this as an outlier. Given the focus on eye medicine, it has a very large number of patients visiting for outpatient procedures at a lower per cost treatments compared to other elective procedures. The remaining 18 reported a combined income from private patients of £195 million over a period of 12 months between 2010–2011.

Those who were able to provide differentiated data indicated that £42 million of their total income was from non-UK resident patients; looking across these 18 hospitals, this meant close to 25% of their private income was from incoming medical tourists. While our sample of hospitals was weighted towards large London-based facilities which do experience a higher number of international patients, income ranged vastly between hospitals surveyed: from over £20million to just £2,466 with a mean of £2.5million.

Those hospitals that were able to report numbers of patients reported a total of 6,722 patients from abroad out of a total of 88,775 private patients counted, i.e. seven percent of private patients were inbound medical tourists. It might therefore appear that medical tourists may be especially profitable, yielding close to a quarter of revenue from only 7% of volume. For a detailed listing of patients and income per hospital, see Annex S2.

DISCUSSION

Results confirm that a small but increasing number of UK patients are travelling abroad to receive medical treatment. Medical travel is complex and not a uniform phenomenon. The majority of UK patients travel within Europe, but an increasing number are seeking treatment further afield. Patients are traveling specifically to Poland and Hungary, and increasingly to India and East Asia. Diaspora, country-specific marketing campaigns, and specific specialism's seem to determine patterns of flows of UK patients seeking care abroad. Patients returning from treatment abroad experience complications.

The analysis demonstrates both the possibility of costs and savings to the NHS as a result of patients travelling abroad, which need to be considered. Unsurprisingly, the largest numbers of inbound medical tourists were in the large hospitals which are internationally known for their specialism; foremost amongst these Great Ormond Street Hospital for Sick Children which reported income of over £20million from 656 patients. Data received and summarised in Appendix 2 also highlights the variation in the percentage of income that international revenue represents for hospitals; to some, especially the large hospitals in London, it marks a significant proportion of private patient income while for others it contributes a very small percentage of funding.

Our analysis of data suggests that the UK is now a net exporter of medical tourists. While incoming medical tourists may be less likely to declare treatment as primary purpose for their visit to the UK than outbound tourists, data over time clearly shows a greater acceleration in outbound over inbound medical tourists. Despite the variations in numbers of patients visiting different hospitals and in the income per patient, the number of medical tourists was comparatively smaller than the percentage of income generated by them (7% of patients generating close to 25% of private income). These figures suggest that non-UK residents travelling to the UK for medical treatment seek high-end specialist expensive procedures, and may generate substantial revenue. Additional numbers of patients for specialist procedures may also help NHS doctors with surgical learning curves.

The changing destinations of UK travellers and the differing origins of those travelling to the UK show that medical travel is a dynamic phenomenon, which can rapidly increase and change. This highlights the importance of continuous routine monitoring to understand medical tourism and to enable researchers, professionals and policymakers to better consider the costs and benefits of medical tourism to the UK.

UK residents who had travelled abroad reported experiencing complications following their return, which echoed case reports in the literature. While we calculated potential costs of these to the NHS, complications experienced also pose an ethical question. There is currently no guidance or regulation on risk or safety for UK residents who consider travelling abroad for treatment. Potential savings as a result of medical travel, especially evident from bariatric patients here, are noteworthy especially at a time of constrained public resources.

Our findings from NHS Trusts indicates that for those wishing to increase their private income as a result of the income cap being raised foreign private patients may be more attractive than domestic private patients.

While this particular research focused on the impact of medical tourism on the UK NHS, the findings give an indication of possible impact of medical tourism in other countries. They are likely to have particular resonance for other universal public health systems.

Strengths and Weaknesses of the Study

While the study used the most robust data set available to measure volume of medical tourism to the UK, the International Passenger Survey, it has several weaknesses. The IPS only surveys 0.2% of travellers entering and leaving the UK. In addition, inbound figures on medical tourists do not provide information on whether these are accessing treatment in the public or the private sector. Interviews with medical tourists also suggested that not all may identify themselves as travelling for medical purposes. Moreover, costs calculated are based on published literature often drawing on small samples.

Thus, although data and analysis presented here represent the most comprehensive analysis of inbound and outbound medical tourism to date, they clearly identify the significant gap in understanding of this increasingly important phenomenon. The particular strength of the findings here lies in the mixed-methods approach. Authors undertook the first comprehensive analysis of the IPS from a medical tourism perspective. Findings were triangulated by drawing on published literature, and by analysis of interviews with 77 UK medical tourists. Similarly, the cost estimates were developed based on results from interviews, costs reported in the published literature, the IPS data set and freedom of information requests to 28 hospital foundation trusts. Hence, each finding has carefully been considered and based on more than one data source.

Directions for Future Research

The impact of medical tourism warrants better monitoring. Findings demonstrate impact in terms of possible costs and benefits and the highly dynamic nature of the phenomenon means that the absolute numbers presented here could grow rapidly. Only continuous monitoring will allow better understanding and informed policy-making to ensure patient safety.

Estimates of cost presented here mark a first step based on the limited data available. To better understand costs and potential savings of medical tourism requires not only better data on volume of travel but also on the differences in long-term health outcomes between patients who travelled and those having received treatment at home. Further research of comparative outcomes is needed.

This research does not explore the ethical dimensions that are involved in many of the considerations relating to medical tourism, including why patients opt to receive care outside of the UK. While data here represents the economic costs of complications experienced by patients these obviously will have to be considered alongside considerations of patient safety.

CONCLUSIONS

UK medical tourism is a growing phenomenon. To fully understand its implications and guide policy on issues such as NHS global activities and patient safety will require investment in further research and monitoring. Despite existing data limitations it is evident that UK medical tourism is dynamic and changing. Findings indicate costs arise where patients travel abroad and return with complications. Analysis also indicates possible savings in the case of specific procedures especially in future health care and social costs averted. Inbound medical tourists offer potentially high income to NHS hospitals. Results of this research may also be indicative of the impact of medical tourism in other public health systems.

DISCLAIMER

The views and opinions expressed therein are those of the authors and do not necessarily reflect those of the HS&DR Programme, NIHR, NHS or the Department of Health.

AUTHOR CONTRIBUTIONS

Conceived and designed the experiments: JH DH NL RS. Analyzed the data: JH DH RS. Wrote the paper: JH DH NL RS. PI on overall research project: NL. Investigator: RS.

REFERENCES

1. Lunt N, Carrera P (2010) Medical tourism: assessing the evidence on treatment abroad. Maturitas 66: 27–32.

2. Smith R, Lunt N, Hanefeld J (2012) The implications of PIP are more than just cosmetic. The Lancet 379: 1180–1181.

3. Vick L (2012) The perils of cosmetic surgery/medical tourism. Clinical Risk 18: 106–109.

4. Pollock AM, Price D, Roderick P, Treuherz T, McCoy D, et al. (2012) How the Health and Social Care Bill 2011 would end entitlement to comprehensive health care in England. Lancet 379: 387–389.

5. Legido-Quigley H, Passarani I, Knai C, Busse R, Palm W, et al. (2011) Crossborder healthcare in the European Union: clarifying patients' rights. BMJ 342.

6. Pollock AM (2013) Will expansion of the NHS abroad benefit UK patients? No. BMJ 346.

7. Smith R, Martı´nez A´ lvarez M, Chanda R (2011) Medical tourism: A review of the literature and analysis of a role for bi-lateral trade. Health Policy 103: 276– 282.

8. Hanefeld J, Smith DR, Lunt N, Horsfall D, King H (under submission (2012)) What do we know about medical tourism - a review of the literature. Health Policy.

9. Johnson TJ, Garman AN (2010) Impact of medical travel on imports and exports of medical services. Health Policy 98: 171–177.

10. Johnston R, Crooks VA, Snyder J, Kingsbury P (2010) What is known about the effects of medical tourism in destination and departure countries? A scoping review. International Journal for Equity in Health 9: (3 November 2010).

11. Pocock NS, Phua KH (2011) Medical tourism and policy implications for health systems: A conceptual framework from a comparative study of Thailand, Singapore and Malaysia. Globalization and Health 7.

12. Smith M (2009) Divided by a common tongue – 'Access'. Br Dent J 206: 185– 185.

13. Culley L, Hudson N, Rapport F, Blyth E, Norton W, et al. (2011) Crossing borders for fertility treatment: motivations, destinations and outcomes of UK fertility travellers. Human Reproduction 26: 2373–2381.

14. Kentikelenis A, Karanikolos M, Papanicolas I, Basu S, McKee M, et al. (2011) Health effects of financial crisis: omens of a Greek tragedy. Lancet 378: 1457– 1458.

15. McKelvey A, David AL, Shenfield F, Jauniaux ER (2009) The impact of crossborder reproductive care or 'fertility tourism' on NHS maternity services. BJOG: An International Journal of Obstetrics and Gynaecology 116: 1520–1523.

16. Miyagi K, Auberson D, Patel AJ, Malata CM (2012) The unwritten price of cosmetic tourism: An observational study and cost analysis. Journal of Plastic, Reconstructive & Aesthetic Surgery 65: 22–28.

17. O'Neil P (2010) Shedding the Pounds: Obesity Management, NICE Guidance and Bariatric Surgery in England. London: Office of Health Economics.

18. Hawkins S, Osborne A, Finlay I, Alagaratnam S, Edmond J, et al. (2007) Paid Work Increases and State Benefit Claims Decrease after Bariatric Surgery. Obesity Surgery 17: 434–437.

4

CHAPTER

Medical Tourism and Policy Implications for Health Systems: A Conceptual Framework from a Comparative Study of Thailand, Singapore and Malaysia

Nicola S Pocock* and Kai Hong Phua

Lee Kuan Yew School of Public Policy, National University of Singapore, 469C Bukit Timah Road, OTH Building, Singapore 259772, Singapore

ABSTRACT

Medical tourism is a growing phenomenon with policy implications for health systems, particularly of destination countries. Private actors and governments in Southeast Asia are promoting the medical tourist industry, but the potential impact on health systems, particularly in terms of equity in access and availability for local consumers, is unclear. This article presents a conceptual framework that outlines the policy implications of medical tourism's growth for health systems, drawing on the cases of Thailand, Singapore and Malaysia, three regional hubs for medical tourism, via an extensive review of academic and grey literature. Variables for further analysis of the potential impact of medical tourism on health systems are also identified. The framework can provide a basis for empirical, in country studies weighing the benefits and disadvantages of medical tourism for health systems. The policy implications described are of particular relevance for policymakers and industry practitioners in other Southeast Asian countries with similar health systems where governments have expressed interest in facilitating the growth of the medical tourist industry. This article calls for a universal definition of medical tourism and medical tourists to be enunciated, as well as concerted data collection efforts, to be undertaken prior to any meaningful empirical analysis of medical tourism's impact on health systems.

INTRODUCTION

Growing demand for health services is a global phenomenon, linked to economic development that generates rising incomes and education. Demographic change, especially population ageing and older people's requirements for more medical services, coupled with epidemiological change, i.e. rising incidence of chronic conditions, also fuel demand for more and better health services. Waiting times and/or the increasing cost of health services at home, coupled with the availability of cheaper alternatives in developing countries, has lead new healthcare consumers, or medical tourists, to seek treatment overseas [1]. The correspondent growth in the global health service sector reflects this demand. The globalisation of healthcare is marked by increasing international trade in health products and services, strikingly via cross border patient flows.

In Southeast Asia, the health sector is expanding rapidly, attributable to rapid growth of the private sector and notably, medical tourism, which is emerging as a lucrative business opportunity. Countries here are capitalising on their popularity as tourist destinations by combining high quality medical services at competitive prices with tourist packages. Some countries are establishing comparative advantages in service provision based on their health system's organizational structure (table 1). Thailand has established a niche for cosmetic surgery and sex change operations, whilst Singapore is attracting patients at the high end of the market for advanced treatments like cardiovascular, neurological surgery and stem cell therapy [2]. In Singapore, Malaysia and Thailand alone, an estimated 2 million medical travellers visited in 2006 - 7, earning these countries over US$ 3 billion in treatment costs (table 2).

Table 1. Health systems in comparison [3]

Country	Thailand	Malaysia	Singapore
Organizational structure	Pockets of excellence in some private Bangkok hospitals	Growing private health sector with movement of qualified workforce	Balanced public-private mix, corporatized public sector
National strategy	Regional health hub Extensive tourism infrastructure	Industrial strategy to develop tourism	Economic growth strategy to develop biomedical industries Regional service hub Medical R&D support
Policy impact	Issues of growing inequity and urban-rural divide	Public-private divide Racial inequities between public and private sectors	Narrow income gaps of public and private sectors

Table 2. Export of health services [2, 4, 5]

	Estimated earnings	No. foreign patients	Origin of patients (in order of volume)	Specialty
Thailand (2006)	Baht 36 billion (US$ 1.1 billion)	1.4 million	Japan, USA, South Asia, UK, Middle East, ASEAN countries	Cosmetic and sex change surgery
Singapore (2007)	S$ 1.7 billion (US$ 1.2 billion)	571 000	Indonesia, Malaysia, Middle East	Cardiac and neuro surgery, joint replacements, liver transplants
Malaysia (2007)	253.84 million MYR (US$78 million)	341 288	Indonesia, Singapore, Japan, India, Europe	Cardiac and cosmetic surgery

Carrera and Bridges (2006) define medical tourism as "the organized travel outside one's natural healthcare jurisdiction for the enhancement or restoration of the individual's health through medical intervention", using but not limited to invasive technology. The authors define medical tourism as a subset of health tourism, whose broader definition involves "the organized travel outside one's local environment for the maintenance, enhancement or restoration of the individual's wellbeing in mind and body". Importantly, their definition of medical tourism takes into account the territorially bounded nature of health systems, where access to healthcare is often but not always limited to national boundaries [6]. Medical tourism constitutes an individual solution to what is traditionally considered a public (government) concern, health for its citizens, who at the micro level are responding to market incentives by seeking lower cost and/or high quality care overseas that cannot be found at home. These tourists may be uninsured or underinsured. Travelling overseas for medical care has historical roots, previously limited to elites from developing countries to developed ones, when health care was inadequate or unavailable at home. Now however, the direction of medical travel is changing towards developing countries [7], and globalization and increasing acceptance of health services as a market commodity [8] have lead to a new trend; organized medical tourism for fee paying patients, regardless of citizenship, who shop for health services overseas using new information sources, new agents to connect them to providers, and inexpensive air travel to reach destination medical [9].The impact of medical tourism on health systems is as yet unknown due to a dearth of data and empirical analysis of the phenomenon.

Governments are noticeably playing a strong marketing and promotional role in the emerging medical tourism industry. This is a clear trend in Southeast Asia, especially in Thailand, Singapore and Malaysia, the main regional hubs for medical tourism, where medical tourist visas are available and government agencies have been established with the mandate to increase medical tourist inflows [10]. Governments in Indonesia, the Philippines and Vietnam have also expressed interest in promoting the industry. The potential economic benefits of medical tourism make it an attractive option for governments. Medical tourism can contribute to wider economic development, which is strongly correlated with improved population health status as a whole, e.g. increased life expectancy, reduced child mortality rates [11]. Encouraging foreign direct investment in healthcare infrastructure and medical tourist inflows with correspondent revenue can create additional resources for investment in health care [12]. Furthermore, medical tourism may slow or reverse the outmigration of health workers, particularly of specialists [13].

However, health systems in some of these countries face challenges in ensuring basic health service coverage for their own citizens [3]. Two tier healthcare provision has emerged in Malaysia, with private services limited to those who can afford it and public services for the rest of the population [14]. Thailand's public to private health worker brain drain has strained public health provision, especially in rural areas [15, 16]. Trade in medical supplies, organs, pharmaceuticals and health worker migration have dominated policy debates about the impact on health systems in developing countries, including concerns about intellectual property rights and access to affordable drugs, the latest medical technology, and retaining doctors and nurses within the public sector and/or within the country's health system at all. There are growing concerns about the impact of medical tourism on health systems, particularly equity of access for both foreign and local consumers [17]. Inequities at home, either by low quality services and/or inability to pay, prompt people to seek cheaper and high quality care treatment overseas. As Blouin (2010) contends, a policy question that remains unanswered is whether medical tourism can improve the capacity of poor people in developing countries to access health services. She calls for the exploration of policy mechanisms that mitigate the risks associated with medical tourism, whilst harnessing the potential benefits, for local consumers [18].

In the academic literature, conceptual analyses of medical tourism have emerged from a tourism management perspective, analysing supply and demand factors [19, 20, 21, 22], and as a node in the trade in health perspective [10, 23, 24, 25, 26]. Legal literature is beginning to cover patient liability issues when surgery is carried out overseas [27]. Recent work has begun to analyse medical tourism and its potential impact on health systems in specific countries [1, 28, 29]. Yet not all health systems functions are analysed in these accounts. A core concern is whether medical tourism diverts resources from public components of health systems in destination countries [30]. Furthermore, conceptual frameworks in the health systems literature focus on the impact of targeted, vertical interventions in health systems [31]. But medical tourism is a phenomenon rather than an intervention; its policy implications have yet to be considered within the context of a health system.

This paper presents a conceptual framework of medical tourism and policy implications for health systems in Southeast Asia, drawing on the cases of Thailand, Singapore and Malaysia, via an extensive review of the academic and grey literature, as well as insights from health consultancies in the public and private sectors across the region. This framework provides a basis for more detailed country specific studies on the benefits and disadvantages of medical tourism, of special relevance for policymakers and industry practitioners in other Southeast Asian countries with similar health systems where governments have expressed interest in facilitating the growth of the medical tourist industry. Bridging the social science disciplines, the public policy approach to research is a pragmatic one, with the end goal of translating research into useful policy recommendations, in this instance those that optimise the benefits of medical tourism for both foreign and local consumers and mitigate the risks. Research methodology is outlined below, followed by the policy implications of medical tourism for health systems at their governance, delivery, financing, human resources and regulation functions [32, 33]. The conclusion emphasizes the need for concerted data collection efforts and identifies variables for further analysis of medical tourism's potential impact on health systems.

RESEARCH METHODOLOGY

Media reports on the medical tourism industry and participation in regional conferences enabled the researchers to pinpoint Singapore, Thailand and Malaysia as the three main hubs for medical tourism in Southeast Asia for comparative analysis. Broadly, there are four types of comparative health policy analyses. The first constitute descriptive studies, with no hypothesis or testing of explanations on why patterns exist, leaving policy explanations implicit for the reader to gauge. The second include collections of international case studies with some assessment of performance, whilst the third type includes studies employing a common framework for analysis (e.g. privatization). The fourth type of cross national studies are those that show a fundamental theoretical orientation, with a specific theme or question as a focus of analysis (Marmor et al 2005: 341 - 2) [34]. We decided to undertake this fourth type of comparative analysis, in order to generate a conceptual framework that could be usefully employed by policymakers to understand the policy implications of medical tourism on health systems with similar structures. Methods employed focussed on conceptualising rather than describing, where one or more new concepts are developed to explain what is being studied [35]. An inductive, theory building approach [36] is appropriate to examine medical tourism where knowledge is far lacking, especially in relation to health systems.

An initial informal literature scan using the search criteria "medical tourism AND Asia" in google scholar revealed a lack of data and authoritative sources on medical tourism, particularly figures for number of patients and estimated earnings. Academic literature was searched exhaustively in the PubMed and Social Science Research Network databases using the search criteria "medical tourism AND Asia" (92) and "medical travel AND Asia" (806), generating a range of mostly conceptual research. Abstracts were scanned for reference to Thailand, Singapore and Malaysia and/or reference to health systems in general. Additional articles were located using the reference list of selected articles. Study selection was not systematic; no article was omitted but considered in the context of health systems/medical tourism in Asia (43). Articles gathered were then categorised according to content focus (e.g. privatisation of health systems, medical tourism empirical evidence, health and trade nexus). Following categorisation, all articles were analysed to identify medical tourism interaction points across the health system functions, with new material continually brought into the analysis. Concurrent to the theory building process, quantitative data on the nature of health systems in the three study countries were retrieved from official country sources and the World Health Organization. These data were triangulated with the academic literature to validate claims made about the nature of health systems. This data also enabled the researchers to make systematic comparisons between the three country health systems. Following this step, grey literature were searched using the above search criteria in Factiva, a news item database, to provide examples of recent developments in the medical tourist industry in the three study countries. Other grey literature sources included management consultancy research reports, working papers on medical tourism, and medical tourism industry player's statistics and promotional materials. Subsequent to analysis and identification of the conceptual framework, potential policy options were outlined based on the literature and/or innovative examples of comparative health policy responses in the region. We anticipated that the different nature of health systems (e.g. mostly public versus private delivery) would also generate differential policy implications according to local context. In the course of our comparative analysis, we found this to be the case to a large extent; however, medical tourism poses potential risks and benefits regardless of the current nature of a health system. As a phenomenon, it can fundamentally change the nature of health systems themselves without policy intervention (e.g. shift towards a dominantly private hospital sector). Thus, the policy implications described are broadly applicable to health systems in general, but of particular relevance to policymakers and industry practitioners in other Southeast Asian countries where governments have expressed an interest in developing the medical tourist industry.

RESULTS

Governance in separate domains of trade and health

Medical tourism straddles the policy domains of trade and health. Its rise is situated within the rapid growth of trade in health services, driven by increased international mobility of service providers and patients, advances in information technologies and communications, and an expanding private health sector [10]. Trade by definition is international, but health systems (financing, delivery and regulation) remain nationally bounded. Additionally, trade objectives of increased liberalisation, less government intervention and economic growth generally do not emphasize equity, whereas health sector objectives like universal coverage do. Consequently, actors in the trade and health policy spheres tend to have conflicting objectives, and trade and health governance processes remain relatively separate at three levels; the international (World Trade Organisation (WTO) and World Health Organisation (WHO)), regional (Association of South East Asian Nations (ASEAN)) and national (government ministries). Reconciling the aims of economic growth with equitable health service provision and access makes governance of medical tourism within a country's health system challenging at best and contradictory at worst.

At the international level, there are clear tensions between the goals of protecting and promoting health and generating wealth through trade [23]. Trade and health policy negotiations occur in isolation, despite the growing importance of the trade and health nexus at the global level, e.g. extensive health worker migration and cross border consumption of health services (medical tourism) [10, 23]. WTO membership requires adherence to a multitude of legally binding obligations, including removal of tariff and non tariffs barriers on goods and services. The WTO's formal governance architecture is embodied in its legally binding trade agreements and compulsory legal dispute mechanism. These legal apparatus afford it more compliance clout than the WHO, which by contrast is an advocacy organization. The WHO imposes no legal obligations on members, relies on non binding agreements, and has no compulsory dispute mechanism. Thus enforcement capacity in cases of non compliance to WHO agreements is limited [23]. Economic growth and trade considerations are likely to surpass health objectives at the global level when countries face sanctions or legally punitive measures for non compliance with trade agreements. Examples of trade and health policy incoherence include patents on essential medicines and tobacco promotion in developing countries, permitted by trade agreements [37].

Whilst most trade in health services takes place outside the framework of existing trade agreements, whether bilateral or multilateral [25], trade in health services including medical tourism is officially provisioned for under the General Agreement on Trade in Services (GATS). The four modes of supply include; 1. The cross border supply of services (remote service provision, e.g. telemedicine, diagnostics, medical transcriptions), 2. Consumption of services abroad (medical tourism, medical and nursing education for overseas students) 3. Foreign direct investment (e.g. foreign ownership of health facilities) and 4. Movement of health professionals [7]. Countries can choose to make GATs commitments (which legally bind them to open markets under the auspices and protection of the WTO) sectorally or via a specific mode. In ASEAN, only Cambodia, Malaysia and Vietnam have made GATs commitments relevant to the health sector [38]. Medical tourism is becoming bureaucratized, formalized and normalized [17] evidenced by GATs provisions for the health sector. In the context of increasing cross border trade in health services, governments have the option to either schedule GATs commitments in health or continue to trade outside of formal agreements. With rapidly changing domestic and international health markets, the latter looks likely, but it is worth noting that GATS commitments can also limit the degree to which foreign providers can operate in the market [39]. In policy terms, this clause can protect health systems from monopolization by foreign investors in the health sector.

Regionally, trade also tends to trump health in terms of policy action. ASEAN is primarily a trade forum, and the 1995 ASEAN Framework on Agreement on Trade in Services (AFAS) makes provisions for services liberalisation between members beyond the WTO GATs. Unlike the WTO, ASEAN has no legal authority to enforce compliance, but a dispute settlement mechanism was recently signed. Whilst the health sector is not covered under the AFAS, it is envisioned that the free flow of all goods, services, investments, capital and skilled labour will be achieved to create an ASEAN Economic Community (AEC) by 2020 [40, 41]. The ASEAN Economic Community (AEC) council meets bi annually to work towards deepening and broadening regional economic integration. In contrast, the ASEAN Health Minister's Meeting (AHMM) is held every two years. Currently, ASEAN health cooperation is limited to disaster preparedness for natural disasters and infectious disease outbreaks. Agreements in health are limited to sanitary and phytosanitary measures, bar a non legally binding Mutual Recognition Agreement (MRA) on the movement of health professionals. The ASEAN Work Plan on Health Development (2010 - 2015) was finalised in July 2010 to cover broader regional health issues, including non communicable diseases, maternal and child health and primary health care [42, 43]. Despite ASEAN's regional economic and health integration, there have been no agreements signed concerning the medical tourism industry. Foreign direct investment by regional players in neighbouring countries is accelerating, with private companies like Singapore's Parkway Holdings (one of the largest hospital operators in Asia) and the Raffles medical group acquiring hospitals in Singapore, Malaysia, Brunei, India and China [26]. Malaysia's state investment company Khazanah's $2.6 billion bid in Parkway Holdings in 2010 gave it a 95% stake in the company [44]. Foreign investment by both private and state investment companies implies that significant profits can be made in the health sector of other countries, with profits accruing to shareholders overseas and few benefits for local consumers, unless profits are taxed and reinvested in the destination health system. The substantive economic capacity of these regional players means that health policy aims, like universal access to healthcare, are likely to come secondary to trade policy aims, like increasing foreign investment that can be gained from medical tourism.

Trade and health policy incoherence in promoting both medical tourism and universal coverage for local consumers at the national level is evident. Whilst several studies on medical tourism allude to government's role in promoting medical tourism [8, 16, 21], these do not differentiate between the role of *different* government ministries and their respective policy aims. Trade and tourism ministries are primarily concerned with increasing economic growth and facilitating international trade in the services sector. In contrast, a health ministry's aim is to improve overall population health and ensure equity in health service access and delivery. Health systems are also nationally bounded;

maximising scarce public resources for health within given territorial constraints gives rise to healthcare protectionism by governments, typified by strict eligibility requirements for access to state subsidised services by migrants. Whilst expansionist medical tourism policies had been initiated in trade and tourism ministries of all three countries, there appears to be a spill over effect on ministries of health (MOH). Increasingly, MOH's are establishing medical tourism committees and departments, dedicated to the promotion of their respective countries' health facilities to other governments/foreign patients. For example, Thailand's medical hub policy was initiated in 2003 by the government agency the Thailand Board of Investment, whilst the Ministries of Commerce, Department of Export Promotion and the MOH in collaboration with private hospitals are now the main implementers of the policy [15]. Whilst Malaysia's national health plan does not mention medical tourism as a strategic aim [45], the MOH formed an inter-ministerial committee for the promotion of medical and health tourism (MNCPHT) in 2003 [28]. Of the three countries, Singapore's government agencies have the most integrated policy stances that strongly support medical tourism [2], reflective of the country's prioritisation of economic growth. Singapore's Tourism Board, the Ministry of Trade and Industry's Economic Development Board and the MOH have set a target to attract 1 million foreign patients by 2012 [46], whilst one of the MOH's explicit priorities is to "exploit the (country's) economic value as a regional medical hub" [47]. In 2004, a multiagency government initiative (including the MOH) SingaporeMedicine was launched with the aim of developing Singapore as a medical hub. Whilst trade and tourism and health ministry objectives are not easily reconciled, medical tourism growth provides an opportunity for inter ministry policy coordination, e.g. via a cross subsidization mechanism whereby medical tourist revenues are taxed, providing extra income for public hospitals. In the three countries, an apparent convergence in trade, tourism and health ministry priorities is taking place, reflective of growing acceptance of health as a private good globally. Improved data collection on medical tourist flows and health systems use and access by local consumers are necessary to assess whether policies that promote medical tourism and universal coverage are reconcilable. Pre-emptively, government ministries should work towards more integrated governance of medical tourism, especially given the highly privatised health system landscape and existing inequities in health systems use and access by local consumers, which could be aggravated by foreign patient inflows.

Delivery in private versus public sector

Medical tourism is driven by the for profit private sector in health systems. The private sector dominates primary care provision in Singapore and Malaysia, but is slowly expanding its role in tertiary hospital care. Private primary care providers are concentrated in urban areas, with public primary care providers catering to those in rural areas, as seen in Thailand and Malaysia [14, 48]. Hospital services are dominated by the public sector, with a 70 - 80% share of beds (table 3) but private hospital providers are steadily growing. In Thailand, private hospital numbers have hovered consistently a 30% of total hospitals between 1994 and 2006 [48]. In Singapore, private sector hospital growth has risen in proportion with public sector hospital growth between 1998 and 2008 [49]. Private hospitals are smaller in size and tend to be located in urban areas, serving middle to high income patients as well as foreign patients [50]. In general, the public private mix of healthcare provision in this region reflects the country's level of economic development. During economic growth periods, wealthier populations have emerged with demand for private providers in response to perceived lower quality public provision. Consequently the public sector has become more pro poor as this group cannot afford private care, leading to the development of a two tier healthcare system seen in Thailand and Malaysia [14, 51]. Public services are generally perceived to be of low quality or unresponsive in this region by local consumers [5052]. The steady growth of private sector hospitals has mirrored the increase in medical tourism (tables 2 and 3).

The link between a growing private, for profit sector that caters to medical tourists and access to such services by local consumers without the ability to pay is elusive. Private ownership of health facilities means that benefits accrued (profits from service fees for foreign patients) are remitted offshore to companies based in different countries who are investing in private hospital chains across Southeast Asia. For example, the recent Fortis-Parkway merger of the second largest Indian healthcare group with the largest private Singapore-Malaysia group created the largest hospital chain in Asia. Parkway's subsequent take-over bid by Malaysia's state investment company Khazanah, means that profits accrued are remitted to Malaysia for health services rendered in Singapore and India. Purchase of costly technology that doesn't have a wider social benefit for the procedures that medical tourists demand has raised concerns about "crowding out" local consumption of high technology procedures [12]. Furthermore, government subsidies for private sector growth, via tax breaks and preferential access to land, is

Table 3. Public versus private health provision [49, 53, 54, 55]

	Hospitals		Beds		Beds per 1000 population	Primary care clinics	
	Public (%)	Private (%)	Public (%)	Private (%)		Public	Private
Thailand	67.9% (2007)	32.1% (2006)	69.3% (2006)	30.7% (2006)	2.2 (2002)	80.5% (2007)	19.5% (2006)
Singapore	63.6% (2009)	36.4% (2009)	80.6% (2009)	19.4% (2009)	3.2 (2007)	1.5% (2005)	98.5% (2005)
Malaysia	40.6% (2008)	59.4% (2008)	77.9% (2008)	22.1% (2008)	1.8 (2007)	32.1% (2008)	67.9% (2008)

unlikely to benefit the health system at large nor facilitate broader public health goals (universal coverage) if private hospitals cater to larger shares of fee paying, foreign patients. This can be seen in Malaysia, where tax incentives are available for building hospitals (industry building allowance), using medical equipment, staff training and service promotion (deductions on expenses incurred) [8]. Private sector growth in health is implicitly encouraged via these benefits, at the same time as government construction of new hospitals has stalled due to alleged insufficient public funds [56].

Medical tourism is emerging in public sector hospitals at the same time as it is being driven by the private sector, notably in corporatized (public) hospitals. Corporatization of hospitals in Singapore since 1985 granted hospitals greater autonomy and exposure to market competition under government ownership, with the aim of lowering costs and improving service quality [57]. All public hospitals in Singapore are Joint Commission International (JCI) accredited [58]. Given that these hospitals are publicly owned, revenues accruing to medical tourism are taxable and thus profits can be reinvested back into the public health system by the government. In Malaysia and Thailand, some public hospitals are allowing their surgeons to operate a private wing for private patients, including medical tourists. This policy move could incentivise surgeons to treat the additional fee paying foreign patients over local consumers, when public health resources are already strained in those countries.

The majority of medical tourists in Southeast Asia hail from neighbouring countries, reflecting inequities in service provision at home, either via unavailability of quality services or underinsurance. In Singapore and Malaysia, most medical tourists are from ASEAN countries, whilst Thailand's consumers are often from outside the region, with the Japanese accounting for the largest share of foreign patients (table 2) [50]. Indonesians travel to Singapore and Malaysia for medical treatment, whilst Cambodians cross the border to Vietnam for higher quality health services. Low quality public and private health provision at home forces them to leave for overseas treatment. Cost is a factor, but Malaysian, Singaporean and Thai hospitals offer specialised services unavailable in other, especially poorer, ASEAN countries [2, 50]. The policy implications go beyond the potential to crowd out consumption by locals. As Chee (2010) points out, when middle class fee paying patients decide to undertake treatment abroad, their domestic health systems lose out, not only financially but in terms of the political pressure that these potential consumers could exert to improve the health system that poorer consumers rely upon [28]. The possibility to "exit" low quality health systems gives the middle class little incentive to exert pressure for quality improvement [59]. Policy options that raise quality standards and minimize quality differentials, both within and between countries in Southeast Asia, would benefit both foreign and local consumers. These include public private linkages via professional exchanges, joint training initiatives, shared use of facilities between public and private providers to maximise resource use, telemedicine, and use of complementary/specialised treatments [1, 12].

Healthcare financing and consumerism

Consumer driven healthcare is becoming the normalised globally and in this region, partly encouraged by governments and the private sector seeking to shift responsibility for one's health to the individual in response to rising healthcare costs and demand for services. Singapore and Malaysia exemplify this trend, as public health expenditure has slowly been declining whilst private health expenditure has increased [28]. The Thai government spent almost double the amount on health as a percentage of total government expenditure (14.1%) compared to Singapore (8.2%) and Malaysia (6.9%) in 2008 [53]. As table 4 shows, the Thai government contributes the majority of total health spending (75.1%), in contrast to Malaysia and Singapore, where private health spending surpasses government health spending. Although both Singapore and Malaysia in theory offer 100% population coverage, high out of pocket payments (OPPs) suggest effective coverage is less than this [52]. Both countries are encouraging greater use of individual financing instruments to pay providers, in addition to compulsory state insurance schemes (Medishield in Singapore) or taxation (Malaysia). These include medical savings accounts (Medisave in Singapore, Employee Provident Fund Account 2 in Malaysia) [60] and widespread private insurance. Thailand is the exception, where the government's commitment to enrolling the population in its universal social insurance scheme means that government investment in health has risen since 2002 [56, 61, 62].

The most regressive financing mechanism, out of pocket payments (OPPs), dominates private health spending in all three countries. More OPPs for services leads to more competition in private healthcare markets, as providers are more likely to compete for patients based on price, especially given the price transparency made possible by the internet. Medical tourist payments are dominated by OPPs, but these payments are becoming more organized as part of insurance coverage.

Table 4. Health expenditure [53]

	Total health expenditure as % of Gross Domestic Product (2008)	Government expenditure on health as % of total government expenditure (2008)	Government health expenditure as % of total health expenditure (2008)	Private expenditure as a % of total health expenditure (2008)	Out of pocket expenditure as a % of private health expenditure (2008)	Private prepaid plans as a % of private health expenditure (2008)
Thailand	4.0%	14.1%	75.1%	24.9%	71.1%	20.9%
Singapore	3.4%	8.2%	35.0%	65.0%	93.9%	2.8%
Malaysia	4.3%	6.9%	44.1%	55.9%	73.2%	14.4%

For example, since March 2010 Singapore's Medisave can be used for elective hospitalizations and day surgeries in hospitals of two partner providers in Malaysia, Health Management International and Parkway Holdings [63]. Deloitte's 2009 medical tourism industry report highlighted four US health insurers who are piloting health plans that permit reimbursement of elective procedure overseas in Thailand, India and Mexico [64]. The trend of insurance companies and employers turning to foreign medical providers to reduce costs looks set to continue as the medical tourism industry grows [29].

One policy implication of the increase in medical tourists on health financing is that differential pricing for foreign patients could drive up costs of services for local consumers over time. Redistributive financing mechanisms may offset these increases. Policy options include taxing medical tourist revenues to be reinvested in the public health system [12], expanding financing instruments that do not tie access to ability to pay (taxation, social insurance) and mandating private providers to participate in schemes that provide coverage to local consumers. Private hospitals could provide services to a specified percentage of foreign patients and local consumers enrolled in state schemes, or provide certain specialist treatment for locals (depending on a centre's area of clinical expertise). The need for such policies is pressing when, for example, private hospitals treating foreign patients in Thailand currently do not participate in social health insurance schemes, which covered 98% of the population in 2009 [25, 52, 65].

Human resources and specialists

Health worker shortages persist to varying degrees in Southeast Asia, at the same time as demand for health services from foreign patients is rising. Whilst all three countries have health worker densities above the WHO critical threshold of 2.28 health workers per 1000 population, all countries face pressures to supply trained health workers to meet population health needs [66, 67]. There are low doctor-to-patient ratios in Thailand and Malaysia (table 5), as well as continual outmigration of doctors from Singapore and Malaysia. Within ASEAN, these two countries record the highest levels of doctor outmigration to OECD countries [68]. International outmigration from Thailand is low, but intra-country migration from rural to urban areas and maldistribution of health workers is common [15, 16]. In response to shortages, Singapore has been able to attract health workers from the Philippines and Malaysia. In Thailand, health workers must pass medical exams in Thai, limiting potential for physician immigration to the country. Whilst

Table 5. Human resources for health [49, 53, 69]

	Doctors per 1000 population	Doctors		Nurses per 1000 population	Nurses	
		Public (%)	Private (%)		Public (%)	Private (%)
Thailand	0.4 (2000)	78.4% (2005)	21.6% (2005)	2.8 (2000)	87.8% (2005)	12.2% (2005)
Singapore	1.5 (2003)	54.8% (2009)	45.2% (2009)	4.5 (2003)	68.5% (2009)	31.5% (2009)
Malaysia	0.7 (2002)	60.1% (2008)	39.9% (2008)	1.8 (2002)	71.2% (2008)	28.8% (2008)

the foreign medical workforce inflow to Malaysia has been substantial, this has been insufficient to offset the outflow of Malaysian doctors to other countries [25].

Rising demand for health services in the region has precipitated the growth in private medical and nursing schools across Southeast Asia and correspondent rise in trained health workers. Public and private medical schools in the region are establishing partnerships with reputable universities overseas. Thailand's Mahidol university nursing department has established links with nursing schools in Sweden, Canada, Australia, Korea, the UK and the USA to facilitate student and teaching exchanges. Singapore's National University recently opened a graduate medical school with Duke university in the USA, and Malaysia's Sunway university medical school trains students in partnership with Monash university in Australia. Such partnerships facilitate capacity building in human resources for health, as well as access to new markets for universities overseas. Importantly, these partnerships signal*quality* of human resources, crucial to the promotion of medical tourism [17].

Developing the medical tourism industry can be seen as a tactic to reduce international emigration of health workers, particularly of specialists. Anecdotal evidence from Thailand indicates that medical graduates, having acquired specialised medical degrees abroad, are finding it lucrative and more satisfying to stay in their home country [2]. Politicians in Singapore have reasoned that in order to recruit and retain specialists in a country with a small local population, that the country must attract a high volume of medical tourists. However, within countries, the growth of medical tourism may exacerbate public to private sector brain drain, notably of specialists who provide elective surgeries demanded by foreign patients. Whilst the proportion of doctors working in the public sector is higher than in the private sector in medical tourist countries (table 5), dual practice, whereby doctors combine salaried, public sector clinical work with fee for service private clientele [70], is common amongst specialists in Thailand and Malaysia. Retaining public sector specialists has become a challenge with the prospect of higher salaries and lower workloads in the private sector. Singapore has managed to maintain competitive public sector salaries, but in Thailand and Malaysia, with larger public - private pay discrepancies, medical tourism has the potential to further incentivise specialists to shift to the private sector. Evidence from Thailand suggests that medical tourism is not negatively impacting the health system by pulling doctors from rural areas. Rather, specialists from teaching hospitals in urban areas are shifting to private hospitals catering to foreign patients [67,71]. All three countries have a high number of doctors with specialty training e.g. 77.5% in Thailand in 2006, [48]. But these specialists are concentrated in the private sector; in Malaysia, only 25 - 30% of specialists work in the public sector [72]. Singapore is the exception, where 65% of specialists are in the public sector [73]. The type of surgery matters; for local consumers seeking specialist, essential surgery (e.g. cardiac, transplantation procedures), paying to see a specialist in a private hospital may be the only option. High quality, specialised care is typically provided in private hospitals and can only be afforded by middle to high income patients [50].

Medical tourism could exacerbate already endemic public to private brain drain in the region. A related concern in Thailand is that medical education is largely publicly funded; private hospitals do not share the costs of such education, yet hire from the same pool of graduates as the public sector [50]. Policy options to mitigate internal brain drain include instituting capitation payments for health costs and standard fees for doctors, regardless of whether a patient is local or foreign. Offering higher salaries in the public sector and bonding publicly funded graduates are options for governments (all three countries bond their graduates for between 3 to 5 years). Dual practice of specialists could be allowed but regulated, so that specialists dedicate a specified amount of time to treat local consumers. When public funds are used to train specialists who then shift to the private sector (potentially to treat medical tourists), redistributive government regulations like paying a fee to leave the public sector (Thailand) may plug a short term financial resource gap, but recruitment and retention is a persistent problem in this region.

Regulation of quality control and new actors

Private hospitals in the three countries are accredited via different channels, leading to differing quality standards between public and private hospitals. Private hospital associations encourage industry self regulation, whereas public hospitals are regulated by the MOH or quasi governmental bodies. For example, publicly owned corporatized hospitals in Singapore operate with autonomy in a competitive environment, but government ownership allows them to shape hospital behaviour without cumbersome regulation [74].

Joint Commission International (JCI) is the most established medical tourist industry accreditor worldwide. Of the three profiled countries,

Singapore has the highest number of JCI accredited providers (18), followed by Thailand (13) and Malaysia (7) [58]. JCI accreditation is an important quality signal to attract medical tourists, but this process is voluntary. The differing quality accreditation channels at the national (private hospital associations vs. MOH) and international levels may lead to inequitable quality standards between the public and private sectors, whereby private hospital standards surpass those in public hospitals, reflective of the current situation in low to middle income countries in Southeast Asia. This has implications for the quality of care received by local consumers without the ability to pay for private services, and the potential divergence of health outcomes between private fee paying patients (foreign and local) and those that can't afford such services. Malaysia's Society for Quality in Health (MSQH), a joint regulatory body launched by the Ministry of Health, Association of Private Hospitals of Malaysia and the Malaysian Medical Association, was recently awarded international accreditation by the ISQua on par with JCI. As the MSQH covers both public and private hospitals, this kind of international standard setting for both sectors could provide a regulatory template for other countries pursuing medical tourism, in order to ensure that both local and foreign consumers enjoy similar quality standards. Policy options include common standards for public and private providers [1] regulated by government, as well as compulsory JCI accreditation for hospitals catering to medical tourists.

New brokers that arise between hospitals and patients are proliferating rapidly. These agencies are located in developed and developing countries, connecting prospective patients to providers via the internet. As yet, the medical brokerage industry has no codes of conduct, and the lack of medical training of brokers raises questions about how these new actors evaluate quality of care when choosing which facilities to promote to prospective patients. There are also no explicit formal standards when establishing referral networks, which could be open to abuse, e.g. financial incentives for brokers from providers to promote facilities) [17]. Regulating medical tourist brokers should be a policy priority in both source and destination countries.

DISCUSSION AND DIRECTIONS FOR FUTURE RESEARCH

Based on the health systems functions of governance, delivery, financing, human resources and regulation [32,33], the conceptual framework (Figure 1) aims to provide a basis for further empirical studies weighing the benefits and disadvantages of medical tourism for health systems, of particular relevance to countries in Southeast Asia.

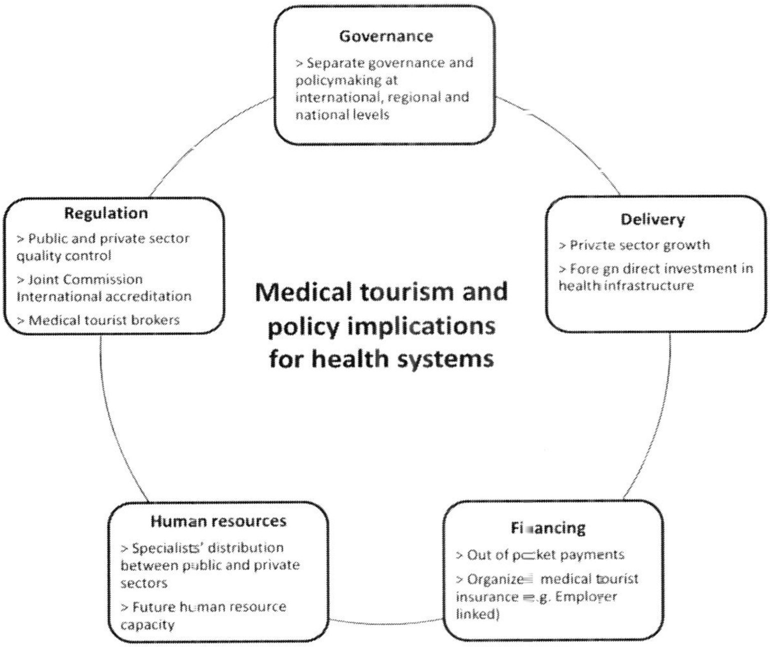

Figure 1. Conceptual framework for medical tourism and policy implications for health systems.

The framework facilitated the identification of the following variables for empirical analysis:

Governance: the number and content of GATs health sector commitments, the number and size of medical tourist government committees or agencies, availability of medical tourist visa.

Delivery: number of hospitals in public and private sector treating foreign patients, consumption of health services by domestic and foreign population (hospital admissions).

Financing: medical tourist revenues, type of medical tourist payment (service fee or insurance, level of copayment), foreign direct investment in the health sector.

Human resources: doctor and nurse ratios per 1000 population, proportion of specialists in the public and private sectors, number of specialists treating foreign patients.

Regulation: number of JCI accredited hospitals, number of medical tourist visits facilitated by brokers.

At present there is an acute lack of reliable empirical data concerning medical tourist flows. Most urgently, a universal definition of who counts as a medical tourist (e.g. per procedure or per inpatient) should be agreed on, ideally at the international (WHO) or regional level (amongst Ministries of Health, Trade, Tourism and private hospital associations). Variation in definitions and estimates amongst the three study countries alone are significant. Singapore's Tourism Board estimates medical tourist inflows based on tourist exit interviews with a small sample population, whilst the Association of Private Hospitals in Malaysia collects data only from member hospitals and includes all foreign patients, including foreign residents and those who happen to require medical care whilst on vacation [28]. Thailand's Ministry of Commerce collects data on medical tourist inflows from private hospitals, counting foreign patients as the APHM does, except that definitions between hospitals about numbers vary (some count inpatient admissions, others per procedure) [71]. Standardised data collection will enable researchers to make meaningful cross country comparisons, as well as carry out detailed country specific studies to investigate the benefits and disadvantages of medical tourism's impact on health systems.

CONCLUSION

The rise of medical tourism in Thailand, Singapore and Malaysia and governments' endorsement of the trend has raised concerns about its potential impact on health systems, namely the exacerbation of existing inequitable resource distribution between the public and private sectors. Nowhere is this more evident than in Southeast Asia, where regulation and corrective policy measures have not kept pace with rapid private sector growth during the past few decades. This paper presents a conceptual framework (Figure 1) that identifies the policy implications of medical tourism for health systems, from a comparative analysis of Thailand, Singapore and Malaysia. This framework can provide a basis for more detailed country specific studies, of particular use for policymakers and industry practitioners in other Southeast Asian countries where governments have expressed an interest in facilitating the development of the industry. Medical tourism can bring economic benefits to countries, including additional resources for investment in healthcare. However, unless properly managed and regulated on the policy side, the financial benefits of medical tourism for health systems may come at the expense of access to and use of health services by local consumers. Governments and industry players would do well to remember that health is wealth for both foreign and local populations.

ACKNOWLEDGEMENTS

All analysis and opinions expressed in this paper are the authors' alone. The authors acknowledge the insights on methodology provided by Wu Xun, as well as the detailed and helpful comments from Chee Heng Leng on an earlier version of this draft.

COMPETING INTERESTS

The authors declare that they have no competing interests.

AUTHORS' CONTRIBUTIONS

NP conducted the research and wrote the first and final versions of the draft. KHP commented on the first and subsequent drafts. NP revised the final manuscript. Both authors read and approved the final manuscript.

REFERENCES

1. Hazarika I: Medical tourism: its potential impact on the health workforce and health systems in India. Health Policy and Planning. 2010, 25: 248-251.

2. UNESCAP: Medical travel in Asia and the Pacific: challenges and opportunities. 2007, Thailand, United Nations Economic and Social Commission for Asia and the Pacific, Accessed 29-09-2010, [http://www.unescap.org/ESID/hds/lastestadd/MedicalTourismReport09.pdf]

3. Chongsuvivatwong V, Phua KH, Yap MT, Pocock NS, Hashim JH, Chhem R, Wilopo SA, Lopez AD: Health and health-care systems in southeast Asia: diversity and transitions. Lancet. 2011, 377 (9763): 429-37.

4. Tourism Malaysia: Media info: Health tourism in Malaysia. 2008, Communications and Publicity Division, Tourism Malaysia, Accessed 29-9-2010, [http://www.tourism.gov.my/corporate/images/media/features/Health%20Tourism%20Oct%202008.pdf]

5. Singapore Medicine: Media Factsheet: Singapore, more than just a world class healthcare destination. 2010, Singapore Medicine

6. Carrera P, Bridges J: Health and medical tourism: what they mean and imply for health care systems. Health and Ageing. 2006, The Geneva Association, 15:

7. Carrera P, Bridges J: Globalization and healthcare: understanding health and medical tourism. Expert Review of Pharmacoeconomics and Outcomes Research. 2006, 6 (Suppl 4): 447-454.

8. Chee HL: Medical tourism in Malaysia: international movement of healthcare consumers and the commodification of healthcare. Asia Research Institute, National University of Singapore, submitted. 2007, Accessed 04-03-2011

9. Whittaker A, Manderson L, Cartwright E: Patients without borders: understanding medical travel. Medical Anthropology. 2010, 29 (4): 336-343.

10. Smith R, Chanda R, Tangcharoensathien V: Trade in health related services. Lancet. 2009, 373: 593-601.

11. Blouin C, Chopra M, Van Der Hoeven R: Trade and social determinants of health. Lancet. 2009, 373: 502-507.

12. Chanda R: Trade in health services. Bulletin of the World Health Organisation. 2002, 80: 158-163.

13. Hopkins K, Labonte R, Runnels V, Packer C: Medical tourism today: what is the state of existing nowledge?. Journal of Public Health Policy. 2010, 31 (2): 185-198.

14. Chee HL: Ownership, control and contention: challenges for the future of healthcare in Malaysia. Social Science and Medicine. 2008, 66: 2145-2156.

15. Pachanee C, Wibulpolprasert S: Incoherent policies on universal coverage of health insurance and promotion of international trade in health services in Thailand. Health Policy and Planning. 2006, 21 (4): 310-318.

16. Wibulpolprasert S, Pengpaibon P: Integrated strategies to tackle the inequitable distribution of doctors in Thailand: four decades of experience. Human Resources for Health. 2003, 1 (12): Accessed 04-03-2011, [http://www.human-resources-health.com/content/1/1/12]

17. Turner L: First world health care at third world prices: globalization, bioethics and medical tourism. Biosocieties. 2007, 2: 303-325.

18. Blouin C: Trade in health services: can it improve access to health care for poor people?. Global Social Policy. 2010, 10 (3): 293-295.

19. Connell J: Medical tourism: sea, sun sand and ... surgery. Tourism Management. 2006, 27: 1093-1100.

20. Garcia Altes A: The development of health tourism services. Annals of Tourism Research. 2005, 32 (1): 262-66.

21. Henderson JC: Healthcare tourism in Southeast Asia. Tourism Review International. 2004, 7: 111-121.

22. Heung VCS, Kucukusta D, Song H: A conceptual model of medical tourism: implications for future research. Journal of Travel and Tourism Marketing. 2010, 27: 336-251.

23. Fidler DP, Drager N, Lee K: Managing the pursuit of health and wealth: the key challenges. Lancet. 2009, 373: 325-31.

24. Lee K, Sridhar D, Patel M: Bridging the divide: global governance of trade and health. Lancet. 2009, 373: 416-22.

25. Arunanondchai J, Fink C: Trade in health services in the ASEAN region. Health Promotion International. 2007, 21 (Suppl 1): 59-66.

26. Diaz Benavides D: Trade policies and export of health services: a development perspective. Trade in Health Services, Global, Regional and Country Perspectives. 2002, Washington DC: Pan American Health Organization, Program on Public Policy and Health, Division of Health and Human Development, 53-69.

27. Cohen G: Protecting patients with passports: medical tourism and the patient protective argument. Harvard Law School Public Law and Legal Theory Working Paper. 2010, 10-08:

28. Chee HL: Medical tourism and the state in Malaysia and Singapore. Global Social Policy. 2010, 10 (3): 336-357.

29. Forgione D, Smith P: Medical tourism and its impact on the US health care system. J Health Care Finance. 2006, 34 (1): 27-35.

30. Pennings G: Ethics without boundaries: medical tourism. Principles of health care ethics. Edited by: Ashcroft RE, Dawson A, Draper H, McMillan JR. 2007, Chicester: John Wiley and Sons, 505-510. 2

31. Atun R, De Jongh T, Secci F, Ohiri K, Adeyi O: Integration of targeted health interventions into health systems: a conceptual framework for analysis. Health Policy and Planning. 2010, 25: 104-111.

32. Badara S, Evans T, Dybul M, Atun R, Moatti JP, Nishtar S, Wright A, Celletti F, Hsu J, Kim JY, Brugha R, Russell S, Etienne C: An assessment of interactions between global health initiatives and country health systems. Lancet. 2009, 373: 2137-69.

33. Hsiao W: What is a health system? Why should we care?. Harvard School of Public Health, working paper. 2003

34. Marmor T, Freeman R, Okma K: Comparative perspectives and policy learning in the world of health care. Journal of Comparative Policy Analysis. 2005, 7 (4): 331-348.

35. Punch K: Introduction to social research: quantitative and qualitative approaches. 2005, London: Sage, 2

36. Glaser BG, Strauss AL: The discovery of Grounded Theory: strategies for qualitative research. 1967, New York: Adline Publishing

37. Zeigler D: International trade agreements challenge tobacco and alcohol control policies. Drug and Alcohol Review. 2006, 25: 567-579.

38. WTO GATs services database. Accessed 10-2-2011, [http://tsdb.wto.org/default.aspx]

39. WTO GATs introduction. Accessed 10-2-2011, [http://www.wto.org/english/tratop_e/serv_e/gats_factfiction1_e.htm]

40. ASEAN: ASEAN cooperation in trade in services. 2000, ASEAN

41. Hew D, Soesastro H: Realizing the ASEAN economic community by 2020: ISEAS and ASEAN - ISIS approaches. ASEAN Economic Bulletin, Institute of Southeast Asian Studies. 2003

42. ASEAN: ASEAN annual report 2009 - 2010: Bridging markets, connecting peoples. 2010, ASEAN

43. ASEAN: Joint statement of the 10th ASEAN health ministers meeting: "healthy people, healthy ASEAN". Singapore. 2010

44. Huang R: Khazanah controls 95% of healthcare firm Parkway Holdings. 2010, Channel News Asia

45. MOH Malaysia: Strategic plan 2006 - 2010. 2008, Ministry of Health Malaysia

46. MTI Singapore: Developing Singapore as a compelling hub for healthcare services in Asia. 2002, Health Sector Working Group, Ministry of Trade and Industry Singapore

47. MOH Singapore: Roles and priorities. 2008, Ministry of Health Singapore, Accessed 14-7-2010, [http://www.moh.gov.sg/mohcorp/about.aspx?id=116]

48. MOH Thailand: Thailand health profile report 2005 - 2007. 2007, Ministry of Health Thailand, Accessed 14-7-2010, [http://www.moph.go.th/ops/thp/index.php?option=com_content&task=view&id=6&Itemid=2]

49. SingStat: Yearbook of Statistics 2009. 2009, Singapore Department of Statistics

50. Arunanondchai J, Fink C: Trade in health services in the ASEAN region. World Bank Policy Research Working Paper. 2007, 4147:

51. O'Donnell O, Doorslaer E, Rannan-Eliya RP, Somanathan A, Adhikari SR, Harbianto D, Garg CC, Hanvoravongchai P, Huq MN, Karan A, Leung GM, Ng CW, Pande BR, Tin K, Tisayaticom K, Trisnantoro L, Zhang Y, Zhao Y: The incidence of public spending on healthcare: comparative evidence from Asia. World Bank Economic Review. 2007, 21 (1): 93-123.

52. Tangcharoensathien V, Patcharanarumol W, Ir P, Aljunid SM, Mukti AG, Akkhavong K, Banzon E, Huong DB, Thabrany H, Mills A: Health-financing reforms in southeast Asia: challenges in achieving universal coverage. Lancet. 2011, 377 (9768): 863-873.

53. WHO: Global Health Observatory. Accessed 20-01-2011, [http://apps.who.int/ghodata/]

54. MOH Malaysia: Health facts 2008. 2009, Health Informatics Centre, Planning and Development Division, Ministry of Health Malaysia

55. National Statistical Office of Thailand: The 2007 private hospital survey. 2007, Ministry of Information and Community Technology

56. Ramesh M, Xun W: Realigning public and private health care in Southeast Asia. The Pacific Review. 2008, 21 (2): 171-187.

57. Phua KH: Attacking hospital performance on two fronts: network corporatization and financing reforms in Singapore.

Innovations in health service delivery: the corporatization of public hospitals. Edited by: Preker AS, Harding A. 2003, Washington DC: Health, Nutrition and Population Series World Bank, 451-485.

58. Joint Commission International: Accredited organizations. Accessed 20-01-2011, [http://www.jointcommissioninternational.org/jci-accredited-organizations/]

59. Hirschman AO: Exit, Voice, and Loyalty Responses to Decline in Firms, Organizations, and States. 1970, Cambridge, MA: Harvard University Press

60. Ministry of Finance Malaysia, Employees Provident Fund. Accessed 20-01-2011, [http://www.kwsp.gov.my/index.php]

61. Phua KH: Health care financing reforms in the Asia-Pacific region. Public Administration and Policy. 2003, 12 (1): 13-36.

62. Phua KH, Chew AH: Towards a comparative analysis of health system reforms in the Asia - Pacific region. Asia Pacific Journal of Public Health. 2002, 14 (1): 9-16.

63. MOH Singapore: Medisave for approved overseas hospitalization. 2010, MOH press release

64. Deloitte Center for Health Solutions: Medical tourism: update and implications. Deloitte. 2009, Accessed 20 - 1 - 2011, [http://www.deloitte.com/us/medicaltourism

65. Hughes D, Leethongdee S: Universal coverage in the land of smiles: lessons from Thailand's 30 baht health reforms. Health Affairs. 2007, 26: 999-1008.

66. WHO: World health report 2006. 2006, Geneva, Switzerland

67. Kanchanachitra C, Lindelow M, Johnston T, Hanvoravongchai P, Lorenzo FM, Huong NL, Wilopo SA, De la Rosa JF: Human resources for health in southeast Asia shortages, distributional challenges, and international trade in health services. Lancet. 2011, 377 (9767): 769-781.

68. Bhargava A, Docquier F, Moullan Y: New medical brain drain data set. 2010, Accessed 10-2-2011, [http://perso.uclouvain.be/frederic.docquier/oxlight.htm]

69. WHO: World health statistics 2009. 2009, World Health Organisation, Accessed 29-9-2010, [http://www.who.int/whosis/whostat/2009/en/index.html]

70. Ferrinho P, Van Lerberghe WV, Fronteira I, Hipolito F, Biscaia A: Dual practice in the health sector: a review of the evidence. Human Resources for Health. 2004, 2: 14

71. Phyu YM, Chotbenjakul W: Medical tourism in Thailand and its impact on the internal brain drain of doctors. Submitted masters thesis. 2010, National University of Singapore, Lee Kuan Yew School of Public Policy

72. Quek D: The Malaysian healthcare system: a review. Intensive workshop on health systems in transition: 29 - 30 April 2009; Kuala Lumpur. 2009, University of Malaya

73. MOH Singapore: Health facts. 2008, Ministry of Health Singapore, Accessed 14-7-2010, [http://www.moh.gov.sg/mohcorp/statistics.aspx?id=5966]

74. Ramesh M: Autonomy and control in public hospital reforms in Singapore. American Review of Public Administration. 2008, 38 (1): 62-79.

5

CHAPTER

Kingdom Of Saudi Arabia: A Potential Destination for Medical Tourism

Shahzad Khan, MD [a,*] and Md. Shariful Alam, PhD [b]

[a] *Wuhan University School of Medicine, Department of Pathophysiology, Wuhan, China*

[b] *School of Business and Economics, United International University, Dhaka, Bangladesh*

ABSTRACT

To perform a comparative study of Medical tourism in the Arab world with special reference to Saudi Arabia and find ways to upgrade medical tourism in Saudi Arabia. A comprehensive literature review and analysis of statistical data from Saudi Ministry of Health is performed.

With more than 37 million health-related trips and the generation of more than £ 33 billion each year, medical tourism has become an important element in the global economy. Travelling abroad to seek medical care is increasing steadily in both developing and developed countries: people in developing countries seek new technology and skills in developed countries, while people in rich developed countries seek medical care elsewhere because of high costs and long waiting lists in their home countries. India, Malaysia, Singapore and Thailand dominate the Asian market, and the United Arab Emirates, with the construction of the Dubai Health Care City, attracts clients from western, Gulf Cooperation Council, Middle Eastern and North African countries. Kingdom of Saudi Arabia is not one of the most popular countries, even though it has all the basic attributes for a successful medical tourism industry, including modern, well-equipped hospitals, a well-established private health care sector, foreign-trained doctors and specialists and a stable, peaceful environment. It is also home to the main holy places of Islam and an important religious center. Kingdom of Saudi Arabia could take the lead in this market by combining medical tourism with religious tourism, not only for the 1.6 billion (23% of the world's population) Muslims but also for the rest of the world. Recent accreditation of many Saudi hospitals by the Joint Commission International in the United States shows their commitment to promoting medical tourism. If properly managed, medical tourism could open the door for an income generation revolution in Kingdom of Saudi Arabia.

Keywords: Economy; Kingdom of Saudi Arabia; Medical tourism; Middle East; Religious tourism

INTRODUCTION

Medical tourism is the practice of travelling to another country with the purpose of obtaining health care, including elective surgery, dental treatment, reproductive treatment, organ transplant and medical check-ups. It is not the same as "wellness tourism", which involves visiting spas and undertaking homeopathic treatment and traditional therapies.[1] Carrera and Bridges[2] described health tourism as "the organized travel outside one's local environment for the maintenance, enhancement or restoration of an individual's well-being in mind and body".

The reasons for medical tourism include3, 4 and 5: lack of availability of a particular facility in the home country; health care in the home country too expensive[5]; health need or procedure not covered by health insurance in the home country; desire to avoid long waiting lists controlled by the government6, 7 and 8; privacy and confidentiality in a far-off setting (for e.g. plastic or cosmetic surgery, fertility treatment, gender assignment procedures and drug rehabilitation)[9]; lack of availability of a specific diagnostic manoeuvre in the home country, regardless of whether treatment is required; combining tourism with an elective procedure (e.g. in Israel, people can have medical treatment and also visit some of the holiest places in the world)10, 11 and 12; annual or routine medical check-up for healthy people[13]; and screening for specific diseases in cases of a family history.[14]

CHOICE OF A SUITABLE COUNTRY FOR MEDICAL TOURISM

When a person decides to undertake medical tourism, the first step is to select a destination. This depends primarily on the availability of the desired facility, cost and technical excellence. A blind decision to select a more developed country for medical tourism is being replaced by local and regional hubs or centres of medical tourism, which attract people from specific regions. Medical tour operators play an important role in choosing countries and may also persuade people to go to specific hospitals and hotels with which they have business contracts.[7] As far as professional excellence is concerned, the hospital chosen should be assessed and accredited by a dependable independent external accreditation body. In international health care accreditation, the quality of health care providers and programmes is ascertained globally. These bodies certify not only hospital services but also a wide range of health care programmes, including primary care, medical transport and ambulatory care. Accreditation Canada (formerly known as the Canadian Council on Health Services Accreditation) is the oldest of such agencies. Others include the Joint Commission International based in the USA, QHA Trent Accreditation in the United Kingdom, the Australian Council for Healthcare Standards and the Society for International Healthcare Accreditation. Some hospitals have multiple international accreditations in order to attract people from many regions. Many countries are applying for international accreditations in order to boost their medical tourism industry. For example, in 2007, at least 15 hospitals in India and 5 in Thailand had Joint Commission International accreditation, whereas there are now 20 hospitals in India and 18 in Thailand,[15] to meet the growing demand. The accreditation is reviewed every 3 years. Table 1 lists the numbers of hospitals with accreditation by Joint Commission International in August 2012.

It should be noted that affiliation with a reputable agency does not necessarily mean that a country or hospital will receive more clients; this also depends on tourism authorities, who use their own tactics to attract patients, regardless of accreditation, such as by offering full services, including any bookings that patients and their families might require, flights, car services, accommodation, clinic appointments, interpreters and recreational activities. Hospitals may also advertise themselves, and tour operators and hospitals may have mutual agreements. Sometimes, the popularity of a region may influence the selection of destination. For example, although China has few accredited hospitals, it attracts many patients from neighbouring countries such as Pakistan and Viet Nam, where people consider that China is more advanced than their own country. Regional or global impact is important and can sometimes surpass accreditation. For example, Germany has few accredited hospitals but attracts many Gulf nationals to its health centres and specialist hospitals. The main reason is that it has an established reputation for high standards of care and advanced medical treatment.[16]

Table 1. Countries and numbers of hospitals with Joint Commission International accreditation in August 2013.

Country	Number of accredited hospitals
Bahrain	1
Bangladesh	1
China	16
Egypt	3
Germany	3
Indonesia	5
Israel	7
Jordan	9
Kuwait	2
Lebanon	2
Malaysia	6
Qatar	5
Pakistan	1
Republic of Korea	13
Kingdom of Saudi Arabia	42
Singapore	14
Taiwan	12
Thailand	18
Turkey	39
United Arab Emirates	39

ECONOMIC VALUE OF MEDICAL TOURISM

Medical tourism is an important source of revenue: the report of the Tourism Research and Marketing Group shows that 37 million health-related trips are taken each year, generating €33 billion.[17] The economic value is not only in the direct health budget but also in indirect factors, as foreign patients are often accompanied by family members or friends, who are likely to use facilities such as hotels, restaurants, tourist attractions and tour operators. The numbers of patients from richer nations going abroad for medical treatment is substantial; according to a McKinsey report based mainly on US citizens, the figure is between 60 000 and 85 000 annually.[18] The annual number of patients travelling from the United Kingdom for this purpose elsewhere in Europe and internationally is estimated by industry sources to be 50 000.[19]

Regional examples show a similar trend. Currently, India, Singapore and Thailand and, in the Gulf, Jordan and the United Arab Emirates (UAE) are paying greater attention to medical tourism increasing their hold on the market in the region. These countries find significant economic development potential in this emerging field. The Indian, Malaysian, Singaporean and Thai governments have all sought to promote their comparative advantages as medical tourism destinations at large international trade fairs, by advertising in the overseas press and providing official support as part of their economic development and tourism policy.[20] For example, 150 000 medical tourists visited India in 2004, according to the Confederation of Indian Industries, and medical tourism may have brought in US$ 2 billion by the end of 2012.[21]

MEDICAL TOURISM IN THE ARAB REGION

The UAE has achieved notable successes in medical tourism and a documented increase in the number of patients coming from abroad to seek medical treatment. According to newspaper reports, Dubai Health Care City showed a sharp growth in medical tourism during the past 3 years. Of 502 000 patients in 2011, about 15% were medical tourists, compared with 10% of 412 000 patients in 2010 and an estimated 5% of 231 000 patients in 2009. Each month, the City Hospital in Dubai receives 10–15 foreign patients, and Al Noor Hospital in Abu Dhabi receives 15% of patients from abroad. The UAE has focused on becoming the hub among Gulf Cooperation Council countries and tries to divert the flow of foreign patients towards the UAE by continuous, efficient advertisement, introducing legislation to facilitate the visa process, building infrastructure, hiring top-class foreign-trained doctors and staff from all over the world with attractive packages and facilities, and international cooperation with medical tour operators.[22]

Jordan has gained a reputation in the region for its commitment to offering services for medical tourism in an emerging, strong medical services sector.[23] In 2005, nearly all foreign patients coming to Jordan were from the Middle East, with the highest percentages arriving from Yemen (29%), Libya (16%), Palestine (12%), Sudan (7%) and other countries in the Middle East and North Africa, i.e. mainly from neighbouring Arabic-speaking countries. According to Jordan's competitiveness report in 2010, medical tourism brings in a revenue of US$ 1 billion annually, with a steady increase of about 10% more foreign patients each year.[23]

Medical tourism in Kingdom of Saudi Arabia: current situation and future scope

Tourism is Kingdom of Saudi Arabia's second largest source of income after oil, generating approximately US$ 13.8 billion annually, and it is the third largest source of employment. Income from international travel and tourism to Kingdom of Saudi Arabia is predicted to reach US$ 63.7 billion by 2019. In Kingdom of Saudi Arabia, religious tourism is the heart and soul of all tourism. It represents the largest economy in the region, with a national budget of US$ 144 billion and a surplus of US$ 20.4 billion. Currently, Kingdom of Saudi Arabia is investing intensively in its health system, most of the financing coming from Government bodies, which places a continuous strain on the Government. According to WHO,[24] total health expenditure as a percentage of gross domestic product was stable in 2009–2011 (latest data available) at about 4.0%, at a total of 79.795 million Saudi Arabian riyals (Table 2). With 386 hospitals, 54 724 hospital beds and 55 000 physicians, a successful health care infrastructure is well on its way to being fully established.[25] Fast-growing health needs in a changing global economic situation demand new ways of health care financing, and medical tourism is a globally tested solution. The recent trend of obtaining accreditation from the Joint Commission International resulted in more than 40 accredited hospitals by August 2012.[15] Although accreditation is not designed to attract more patients from abroad but rather to upgrade national health facilities, the authors believe that this will develop more trust among foreign patients choosing a hospital. Most doctors in Saudi hospitals have been trained in western countries with well-known, trustworthy training systems, which may attract patients from abroad.

Table 2 .Gross health expenditure in Kingdom of Saudi Arabia, 2011.

Indicator	2011 value
Total health expenditure (% gross domestic product)	4
General Government expenditure on health (% total health expenditure)	69
Private expenditure on health (% total health expenditure)	31
Total expenditure on health per capita at purchasing power parity (riyals per US$)	901
Total expenditure on health (million riyals)	79 795
General government expenditure on health (million riyals)	55 003
Private expenditure on health (million riyals)	24 792

Merging religious and medical tourism

Surveys and research suggest that most foreign patients coming to Kingdom of Saudi Arabia have a spiritual life and regard their spiritual and physical health as equally important. Furthermore, during illness, people may have greater spiritual needs. Several studies have shown a potential impact of religious or spiritual activity on physical and mental health. Most have shown that religious involvement and spirituality are associated with better health outcomes, including faster recovery, greater longevity, better coping skills and feelings of optimism and hope, with reduced feelings of depression and anxiety.26 [and] 27 The fact that Kingdom of Saudi Arabia is home to the most important holy places of Islam makes religious tourism the backbone of Saudi tourism. With proper arrangements, patients (especially those for elective surgery) could be offered medical or surgical treatment packaged with religious tourism, such as visits to holy places for prayers or *umrah*. Israel is a good example, as many medical tourists select Israel as a destination because of its attractive sites, such as the Dead Sea. [13] Moreover, most Muslims believe that God hears their prayers and grants health in the holy places, and they believe that spiritual activities in the holy places boost their healing process. Many old patients with terminal stage diseases or geriatric problems would like to spend time in the holy places during treatment. The spiritual element is an essential component of palliative and end-of-life care.[28]

Kingdom of Saudi Arabia's achievements in health care

Kingdom of Saudi Arabia has made notable achievements in some health care-related fields, which could lead the nation's potential health-related progress. For example Kingdom of Saudi Arabia has a well-developed organ transplant set up and has a good history of successful transplants (Table 3); however, many Saudis seek transplants in other countries, underestimating their own capability. A recent study showed high mortality and morbidity rates in Saudi and Egyptian patients who received transplants in China when compared with those transplanted in King Faisal Specialist Hospital and Research Centre during the same period.[29] As described earlier, Saudi hospitals are equipped with well-trained professionals and the most advanced technology, and this edge could be used to market medical tourism. The facilities include well-equipped critical care units such as for burns, surgical intensive care, endoscopy, neurosurgery and general surgery. All these are attractive reasons for the people of the region to select Kingdom of Saudi Arabia as their destination for meeting their health requirements. Neurological rehabilitation for the handicapped, rehabilitation with both occupational therapy and physiotherapy and rehabilitation for people with neurological disorders due to stroke and cerebral palsy are particularly suited for the Arab population of the region, who require not only professional and technological proficiency but also cultural and language similarities, which can hasten their recovery. Kingdom of Saudi Arabia's capacity can be judged from the fact that, in 1 year, 463 481 cases were treated in Ministry of Health hospitals and 949 876 cases in other Government sector hospitals, with physiotherapy, speech therapy and various types of rehabilitation.[30] Moreover, 353 203 disabled children were successfully rehabilitated, including speech therapy, physiotherapy and psychological consultations.[31] A successful record in surgery, especially for separating Siamese twins,[32] could be a focus for medical tourism.

Table 3. Organs successfully transplanted in Kingdom of Saudi Arabia, 1983–2009.

Transplanted organ	Period	Total
Lung		32
Cornea	1983—2008	18 225
Kidney	1979—2009	6 143
Heart valves	1986—2009	497
Heart	1986—2009	168
Liver	1990—2009	828

Barriers to medical tourism in Kingdom of Saudi Arabia

Underestimation of the country's potential

The authors consider that the main problem in the emergence of Kingdom of Saudi Arabia as a hub of medical tourism is that it is neglected by the Government and the private sector, which underestimate the importance of medical tourism. Major medical tour operators are of the opinion that Kingdom of Saudi Arabia has not made a significant mark in the inbound medical tourism business.[33] The authors suggest that the relevant authorities should make a consolidated, joint effort to build a long-term plan and infrastructure for medical tourism.

Difficult documentation and visa processes

It is difficult to acquire a medical visa for a number of reasons.[33] A delay in acquiring a visa is unfavourable in the case of medical ailments, when one of the first priority of patients is early diagnosis and treatment. This delay cannot be reduced until the Government recognizes the importance of medical tourism and makes special arrangements and policies for this activity.

Insufficient information

Although many milestones have been achieved in the concept of E-health, special attention is required for foreign patients intending to obtain medical care in Kingdom of Saudi Arabia. This includes clear-cut information from the medical unit to which the patient wishes to go. Special information portals should be dedicated solely for this purpose, and a more efficient national health information system[34] should be set up, from which the international community could also benefit.

Poor cooperation between medical tour operators and medical units

Strong partnerships with medical tour operators are of key importance, as these organizations are not only specialized in travel services but can also assist in related services before and throughout a trip. Depending on the agreements with medical units, they can take on most of the burden of scheduling appointments, booking hotels and providing assistance in getting a medical visa.

Inadequate marketing

As Kingdom of Saudi Arabia's neighbours are emerging giants in medical tourism, Kingdom of Saudi Arabia should undertake a well-planned, continuous, worldwide publicity campaign through sources such as the Internet, social media and print and e-media.

SUGGESTED REMEDIES

Identify specific areas for medical tourism

Although providing all types of medical services is a fascinating target, focussing on a few areas first and mastering them would be more commercially practical in establishing a position. For example, Thailand has focused on gender defining and cosmetic surgery, Israel on in-vitro fertilization, China on stem cell treatments, India on alternative medicine, bone-marrow transplant, cardiac bypass surgery and eye surgery.[35] Kingdom of Saudi Arabia must explore which field its hospitals should focus on.

Identify the target population

As mentioned earlier, certain populations are attracted more to certain locations. This might be due to a similar ethnic and racial background: Taiwan attracts mainly Chinese patients, Jordan mainly Arabs from the Middle East and North Africa, and India and Pakistan mainly South-East Asians. Kingdom of Saudi Arabia has the edge, in that it can attract not only patients from Gulf Cooperation Council, Middle Eastern and North African countries but also, if packaged with religious tourism and wisely advertised, the entire Muslim world.

Make policy and legal issues easy and systematic

Legal regulations and laws related to medical tourism, visa issuance, medical tour operators, private sector involvement, health insurance, medico-legal aspects of malpractice and related issues should be revised and clear enough for the convenience and safety of incoming patients. The ministries of health, tourism and information should initiate comprehensive, multi-ministerial efforts.

More international cooperation

Cooperation with foreign governments and private bodies will have dual benefits, such as publicity at the cross-border level and, more importantly, direct access to Saudi health facilities. The Ministry of Health has tried a number of strategies in this regard, including cooperation with the governments of Germany, the Netherlands and Thailand. One possibility is operation of selected medical units or complete hospital systems by well-known foreign health care companies. Bearing in mind the pros and cons of this approaches, the Ministry of Health has instituted a pilot system for 31 public hospitals across the country, which are continuously monitored. Expansion of the policy will depends on the outcome of this study.[34]

Expanded use of Internet-based health information and E-health

e-Health and electronic information systems are already functional in many Saudi hospitals, such as the King Faisal Specialist Hospital and Research Centre, the National Guard Health Affairs and Medical Services of the Armed Forces.[36] The Ministry of Health understands the importance of E-health and undertook a 4 billion-riyal project between 2008 and 2011.[37] To emphasize the importance of e-health and to explore future strategies and infrastructure, a series of conferences on e-health is organized by the Saudi Association for Health Informatics every 2 years (the latest one was in 2012).[38] Yet another success in bringing the benefits of information technology to medical care is the creation of 'telecare' systems to follow up patients with chronic diseases after discharge from hospital. A telecare system is already functioning in the country. For example, the telecare system for managing diabetes at the Saudi National Guard Hospital in Riyadh is a web-based programme that allows diabetes patients to communicate online with their clinic in order to follow their blood sugar levels, without having to visit the clinic.39 [and] 40This system would be of great benefit to foreign clients, who would feel secure and attached to the hospital even after discharge.

CONCLUSION

Kingdom of Saudi Arabia has good potential for becoming a hub of medical tourism in the region. It has most of the infrastructure and facilities with high standards, many of which are already accredited internationally. The unique position of the Kingdom in religious and archaeological history and its political stability give it advantages over competitors in the field. Having learnt lessons from the achievements of neighbouring countries, it is now time that Kingdom of Saudi Arabia become motivated and initiate a well-planned, long-term plan for the medical tourism industry. The current era of global economic turmoil is the right time to take advantage of medical tourism and reduce the heavy dependence of the country on oil.

REFERENCES

1. Smith R, Martı´nez A´ lvarez M, Chanda R. Medical tourism: a review of the literature and analysis of a role for bi-lateral trade. Health Policy 2011; 103: 276e282.

2. Carrera PM, Bridges JF. Globalization and healthcare: understanding health and medical tourism. Expert Rev Pharmacoecon Outcomes Res 2006; 6: 447e454.

3. Burkett L. Medical tourism. J Leg Med 2007; 28: 223e245.

4. Dunn P. Medical tourism takes fight. Hosp Health Netw 2007; 81: 40e42.

5. Milstein A, Smith M. America's new refugeesdseeking affordable surgery offshore. N Engl J Med 2006; 355: 1637e1640.

6. Ramirez de Arellano AB. Patients without borders: the emergence of medical tourism. Int J Health Serv 2007; 37: 193e198.

7. Turner L. Medical tourism: family medicine and international health-related travel. Can Fam Physician 2007; 53: 1639e1641.

8. Snyder J, Crooks VA, Johnston R, Kingsbury P. What do we know about Canadian involvement in medical tourism? A scoping review. Open Med 2011; 5: e139ee148.

9. Horowitz MD, Rosensweig JA, Jones CA. Medical tourism: globalization of the healthcare marketplace. Medscape Gen Med 2007; 9: 33.

10. Sukenik S, Abu-Shakra M, Kudish S, Flusser D. Dead Sea and Tiberias as health resort areas for patients suffering from different types of arthritis. Harefuah 2006; 145: 117e122.

11. Ingber A. Medicine is flourishing at the Dead Sea. Isr Med Assoc J 2006; 8: 499e500.

12. Moses SW, David M, Goldhammer E, Tal A, Sukenik S. The Dead Sea, a unique natural health resort. Isr Med Assoc J 2006; 8: 483e488.

13. O'Hagan HC. Medical tourists opting for sea, sand, sun and surgery. Abu Dhabi: National Abu Dhabi Media Co; 2012.

14. Oh KM, Jun J, Zhou Q, Kreps G. Korean American women's perceptions about physical examinations and cancer screening services offered in Korea: the influences of medical tourism on Korean Americans. J Community Health; 2013 [Epub 2013/12/11].

15. JCI accredited organizations. Oak Brook, Illinois: Joint Commission International; 2013.http://www.jointcommissioninternational. org/jci-accredited-organizations/ [accessed 05.09.13].

16. Connell J. Medical tourism: sea, sun, sand and. surgery. Tourism Manage 2006; 27: 1093e1100.

17. Smith RD, Chanda R, Tangcharoensathien V. Trade in healthrelated services. Lancet 2009; 373: 593e601.

18. Ehrbeck T, Guevara C, Mango PD. Mapping the market for medical travel. McKinsey Q; May 2008.

19. Medical tourism survey. Churchgates, Herts: Intuition Communication Ltd; 2007.http://www.treatmentabroad.com/about/medical-tourism-survey/?locale¼en [accessed 09.03.13].

20. Lunt N, Carrera P. Medical tourism: assessing the evidence on treatment abroad. Maturitas 2010; 66: 27e32.

21. Chinai R, Goswami R. Medical visas mark growth of Indian medical tourism. Bull World Health Organ 2007; 85: 164e165.

22. Ismail M. Dubai Healthcare City to compete for foreign patients. The National. http://www.thenational.ae/news/uae-news/tourism/dubai-healthcare-city-to-compete-for-foreign-patients; 18 February 2012 [accessed December 2013].

23. Jordan's competitiveness report, chapter 6: medical tourism. Amman: Ministry of Planning and International Cooperation; 2007.

24. NHA indicators. Geneva: World Health Organization; 2013.http://apps.who.int/nha/database/DataExplorer.aspx?ws¼0&d¼1 [accessed 01.09.13].

25. Al-Yousuf M, Akerele TM, Al-Mazrou YY. Organization of the Saudi health system. East Mediterr Health J 2002; 8: 645e653.

26. Boelens PA, Reeves RR, Replogle WH, Koenig HG. The effect of prayer on depression and anxiety: maintenance of positive influence one year after prayer intervention. Int J Psychol Med 2012; 43: 85e98.

27. Allen RS, Harris GM, Crowther MR, Oliver JS, Cavanaugh R, Phillips LL. Does religiousness and spirituality moderate the relations between physical and mental health among aging prisoners? Int J Geriatr Psychiatry 2013; 28: 710e717.

28. Hayden D. Spirituality in end-of-life care: attending the person on their journey. Br J Community Nurs 2011; 16: 546.

29. Allam N, Al Saghier M, El Sheikh Y, Al Sofayan M, Khalaf H, Al Sebayel M, et al. Clinical outcomes for Saudi and Egyptian patients receiving deceased donor liver transplantation in China. Am J Transplant 2010; 10: 1834e1841.

30. Usman NAA. Health statistical year book. Riyadh: Ministry of Health; 2009 [Report No. 15/0588].

31. Services of rehabilitation center for disabled children by age. Riyadh: Saudi Society of Caring Handicapped Children; 2009.

32. National guard health affairs. Wikimedia Foundation, Inc; 2013.http://en.wikipedia.org/wiki/National_Guard_Health_Affairs#cite_note-2 [accessed 24.04.13].

33. Davis M. The global emergence of Saudi Arabia in medical tourism. Raleigh, North Carolina: ArabMedicare.com; 2009. http://www.arabmedicare.com/Medical_Tourism/news_Ziyad_Bin_Mahfouz_Interview.htm [accessed 01.03.13].

34. Almalki M, Fitzgerald G, Clark M. Health care system in Saudi Arabia: an overview. East Mediterr Health J 2011; 17: 784e793.

35. Medical tourism in India. Wikimedia Foundation, Inc; 2013.http://en.wikipedia.org/wiki/Medical_tourism_in_India [accessed 01.03.13].

36. Altuwaijri MM. Electronic-health in Saudi Arabia. Just around the corner? Saudi Med J 2008; 29: 171e178.

37. Mufti MH. Healthcare development strategies in the Kingdom of Saudi Arabia. Berlin: Springer; 2000.

38. Riyadh Saudi Association for Health Informatics Conference 2009., http://www.saudiehealth.org/; 2012 [accessed 01.09.13].

39. Telecare for Managing Diabetes TeMaD. Riyadh: Ministry of Health; 2013.http://eapps.ngha.med.sa/newapps/temad/user/index.aspx [accessed 04.09.13].

40. Alkadi K, Roudsari A. TeMaD system: telecare for managing diabetes in

6

CHAPTER

Understanding India, Globalisation and Health Care Systems: A Mapping of Research in the Social Sciences

Ramila Bisht[1]*, Emma Pitchforth[2] and Susan F Murray[3]

[1]*Centre of Social Medicine and Community Health, School of Social Sciences, Jawaharlal Nehru University, New Delhi 110067, India.*

[2]*RAND Europe, Westbrook Centre, Milton Road, Cambridge CB4 1YG, UK.*

[3]*King's College London, James Clerk Maxwell Building, Waterloo Road, London SE1 8WA, UK.*

ABSTRACT

National and transnational health care systems are rapidly evolving with current processes of globalisation. What is the contribution of the social sciences to an understanding of this field? A structured scoping exercise was conducted to identify relevant literature using the lens of India – a 'rising power' with a rapidly expanding healthcare economy. A five step search and analysis method was employed in order to capture as wide a range of material as possible. Documents published in English that met criteria for a social science contribution were included for review. Via electronic bibliographic databases, websites and hand searches conducted in India, 113 relevant articles, books and reports were identified. These were classified according to topic area, publication date, disciplinary perspective, genre, and theoretical and methodological approaches. Topic areas were identified initially through an inductive approach, then rationalised into seven broad themes. Transnational consumption of health services, the transnational healthcare workforce; the production, consumption and trade in specific health-related commodities, and transnational diffusion of ideas and knowledge have all received attention from social scientists in work related to India. Other themes with smaller volumes of work include new global health governance issues and structures; transnational delivery of health services and the transnational movement of capital. Thirteen disciplines were found represented in our review, with social policy being a clear leader, followed by economics and management studies. Overall this survey of India-related work suggests a young and expanding literature, although hampered by inadequacies in global comparative data, and by difficulties in accessing commercially sensitive information. The field would benefit from further cross-fertilisation between disciplines and greater application of explanatory theory. Literatures around stem cell research and health related commodities provide some excellent examples of illuminating social science. Future research agendas on health systems issues need to include innovative empirical work that captures the dynamics of transnational processes and that links macro-level change to fine-grained observations of social life.

Keywords: Social science research Globalisation and healthcare India State of the art literature review Transnational Trade

INTRODUCTION

Current globalisation is regarded as a new phase of world integration with increased density and frequency of international or global social interactions relative to local or national ones. In these new dynamics, nation states are influenced by transnational processes occurring on multiple economic, political and cultural levels [1]. Much public health literature still assumes a default model of a 'healthcare system' bounded by a nation state which to varying degrees takes responsibility for the four essential functions of service provision, resource generation, financing and stewardship [2], but there is increasing awareness that globalisation processes are shaping local health systems in important ways [3]. In this article we set out to map through the lens of India's healthcare and its relationship to transnational health care systems, the current extent and form of the social science approaches to an understanding of this field.

India is an illuminating case for which one would expect to find a considerable body of literature. The country has emerged as one of the so-called 'Rising Powers' which represent drivers of global economic and social change. Healthcare is one of India's largest sectors in terms of revenue and employment, and is expanding rapidly. The private sector accounts for more than 80% of total healthcare spending within India [4]. Since 'liberalisation' of economic policies in the late 1980s, India has also become increasingly attractive to foreign investors with its low costs and large, English-speaking, workforce. During the 1990s, Indian healthcare grew at a compound annual rate of 16% and today the total value of the sector is more than $34 billion and projected to grow to nearly $40 billion by 2012. The consultancy group Pricewaterhouse Coopers' *Emerging Market Report on Healthcare in India*[5] listed a number of 'flourishing market opportunities' for foreign as well as national companies: medical tourism (often combining elective surgery and aspects of Ayurveda), the emerging health insurance market, telemedicine, healthcare infrastructure expansion (including public-private partnerships and training of healthcare staff), the medical equipment market, and the pharmaceuticals industry (manufacture, research and clinical trials). Indian finance capital and healthcare-related firms are expanding horizons to markets and investment opportunities outside its borders. Indian pharmaceutical production, for example, supplies some 22% of the global generics market, primarily exporting to the US and Europe but also to China, Brazil, Nigeria and Mexico. A significant proportion of the vaccines produced are for developing countries [6].

Despite the notable rapid economic growth India is ranked 134 out of 187 countries in the Human Development Index [7]. India currently has great social and economic inequalities. There are marked disparities among different geographical regions, between social groups, among different income levels and between the sexes. The Indian 'middle class' is rapidly growing but over a third of the population live on less than US$1 a day, and around a third of the adult population (34.8%), including over 190 million Indian women remain illiterate [8].

The complex and dynamic relationship between India's adaption to the new global environment and Indian population heath, health systems and healthcare related industries is one deserving of social science scrutiny. In the 1990s, civil society organisations actively campaigned against the healthcare reforms of the structural adjustment policy and produced considerable writings on the negative impact of this process on health and health services. A few landmark studies at that time charted the patterns of utilisation of health services for acute and long term treatment [9, 10, 11] and the costs incurred and its implications for equity [10, 12, 13]. Initially academic research on these themes mostly reflected critical public health interests, at times influenced by social science perspectives. In recent years this has opened out to include an increasingly diverse range of contributions from across the social science disciplines.

This article reports on a scoping exercise conducted to establish the extent of social science research concerning India, globalisation and healthcare systems. Findings are reported where relevant but we seek primarily to bring together an account of the themes, questions, methods and challenges encountered in social science research, and to identify avenues for future work. It hopes to highlight some of the challenges of a health systems and globalisation agenda and of researching the interface between macro level changes and policies and meso and micro level enactment.

Literature review process

We adapted previous scoping review methodology [14] to devise a five step method of scoping and analysis that could incorporate a broad analytic framework relevant to globalisation and health care (Figure 1). In step 1 conducted in the summer of 2010, eleven electronic bibliographic databases; Scopus, International Bibliography of the Social Sciences (IBSS); SocINDEX; Business Source Premier; the PubMed databases (PubMed.gov, PubMed Central and BioMed Central); Applied Social Sciences Index and Abstract (ASSIA); International Political Science Abstracts; Social Sciences Citation Index (Web of Science); Sociological Abstract; Human Resources Abstracts and EconLit were searched. The search terms (see Figure 1) used were derived through a process of refinement. In those databases in which a search combination using the term 'health' led to several thousand articles, the modification 'health care' was used, in others this proved too limiting and 'health OR health care' was employed.

In step 2, by reviewing the title, journal name and abstracts, we selected articles that: a) constituted a social science contribution to the field and; b) were published in English. We employed a liberal definition of what constitutes a 'social science contribution', accepting any of the following - publication in a social science journal; funding received from a social science research council; a first author working in a social science department, faculty or institution; a first author whose discipline was identified as one of the social sciences in the article or whose social science disciplinary background was publically known. The 'social science' disciplinary areas were considered to include all those listed as such by the Economic and Social Research Council in the UK (see Table 1). 'Nursing research' was included where the research fell into a category of education or human resource management, or had received funding from the ESRC or equivalent funder of social science research. The majority of papers excluded at this stage were epidemiology or medical science papers, many of these relating to the global burden of disease. Research in India is not always published in indexed journals and many scholars publish in books rather than journals. The database searches were supplemented at this point by a search of Google Scholar for specific authors or subjects and relevant books and chapters and grey literature known to the authors were added.

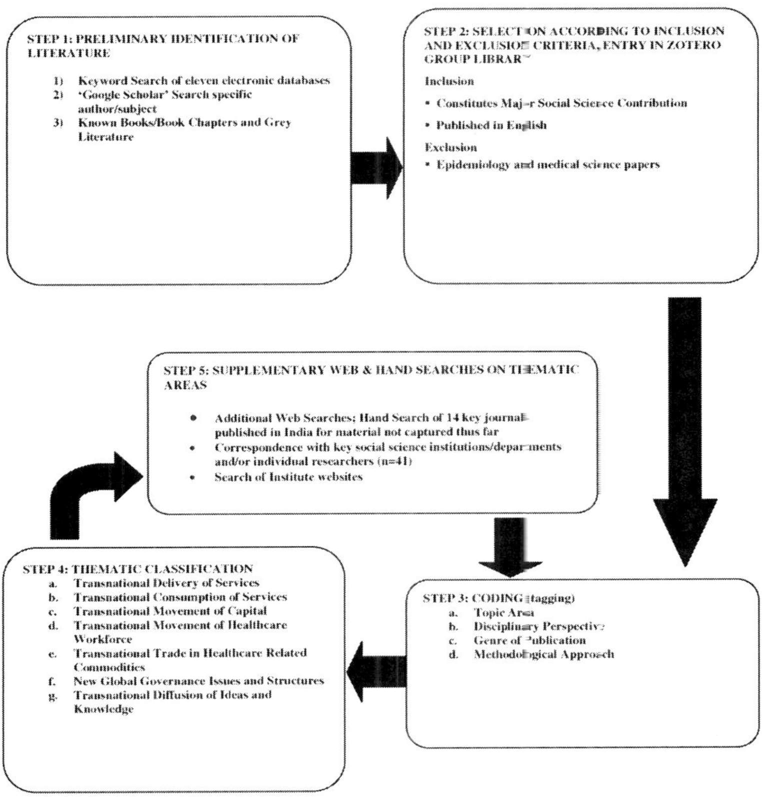

Figure 1. Methodology for Literature Review.

Table 1. The social sciences: 19 disciplines listed by the Economic and Social Research Council (ESRC) of the UK

a)	**Area and Development Studies**
b)	Demography
c)	**Economic and Social History**
d)	**Economics**
e)	**Education**
f)	**Human Geography**
g)	Science and Technology Studies
h)	Linguistics
i)	**Management and Business Studies**
j)	Environmental Planning
k)	**Political Science and International Studies**
l)	Psychology
m)	**Social Anthropology**
n)	**Social Policy**
o)	Social Work
p)	**Socio-Legal Studies**
q)	**Sociology**
r)	Statistics
s)	Methods and Computing

Note: Disciplines represented in this Review are shown in bold.

The 76 relevant references were entered in Zotero bibliographic software group library. In step 3, initial 'tagging' (coding) was undertaken. Articles were classified according to topic area, genre, disciplinary perspective and theoretical and methodological approaches (Figure 1). Topics were initially identified through an inductive approach similar to thematic analysis. These were later rationalised into a number of thematic areas as described below (see Step 4 in Figure 1).

Once the seven thematic areas were identified further web searches and hand searches of bibliographies of key articles were carried out in late 2010 and early 2011 for any material relevant to the themes which had not been captured in the earlier search of the electronic databases (Step 5). Hand searches became a significant element in our methodology as it had become clear that Indian literature is not easily accessible through the electronic databases. A total of 14 journals published from India that were not covered in the electronic databases were hand searched or web-searched. This included *Economic and Political Weekly*, a key journal used by Indian social scientists but not indexed on the above databases. This resulted in 16 additional articles. Finally we wrote to 41 key social science researchers, institutions or departments in India, requesting relevant social science studies and reports on work conducted in the last two decades. Six additional articles were identified in this way, and a search of the websites of these institutions identified 15 further non-duplicate hits. In total 113 articles and chapters were identified as relevant for inclusion in this review.

Findings

The scoping process resulted in 113 relevant journal articles and chapters. We identified seven thematic areas across this literature. These were: (a) transnational movement of workforce; (b) transnational consumption of services; (c) production, consumption and trade in specific health-related commodities; (d) transnational diffusion of ideas and knowledge; (e) new global governance issues and structures; (f) transnational delivery of services; (g) transnational movement of capital. Table 2 provides an overview of papers in each area in terms of number of articles, publication dates and the social science disciplines represented. Some articles covered more than one theme. The table is effective in showing that certain disciplines, such as economics, are represented across most of the themes whilst sociology and others are concentrated in two or three themes. In total there are contributions from 11 out of the 19 social science disciplines listed by ESRC. The work in each of the thematic areas is summarized below. The summaries aim to describe the main disciplinary and methodological approaches, types of paper and research questions. Particular emphasis is given to highlighting empirical studies or explicit use of social science theory.

Table 2. Representation of social science disciplines as listed by the Economic and Social Research Council (ESRC), UK

	Number of articles	Publication date	Area and Development Studies	Economic and Social History	Economics	Education	Human Geography	Management and Business Studies	Political Science and International Studies	Social Anthropology	Social Policy	Socio-Legal Studies	Sociology
Transnational movement of healthcare workforce	29	1989-2009			✓	✓	✓	✓			✓		✓
Transnational consumption of services	21	2004-2011			✓		✓	✓			✓	✓	
Production, consumption and trade in specific healthcare related commodities	21	1998-2010		✓			✓		✓	✓			✓
Transnational diffusion of ideas and knowledge	19	1998-2010	✓					✓	✓	✓	✓		✓
New global governance issues and structures	14	2000-2010			✓			✓	✓			✓	
Transnational delivery of services	8	2001-2009			✓			✓				✓	
Transnational movement of capital	7	2001-2010			✓						✓		

Transnational delivery of services

The economist Rupa Chanda has indicated that the healthcare sector is among the most rapidly growing in the world economy, and globalisation of health services is reflected in the growing cross-border delivery and in increasing numbers of joint ventures [15]. Cross border delivery includes shipment of laboratory samples, diagnosis, and clinical consultation via traditional mail channels as well as the rapidly expanding 'e-health' services such as electronic delivery of health services such as surveillance, and diagnostics including lab testing, tele-radiology, tele-consultations and remote surgery, and indirect services such as teleconferencing and tele-education, medical transcriptions and claims processing [16].

This area has not as yet received much attention from social scientists. Eight articles making reference to India were identified, all published since 2001, when Chanda's working paper on 'Trade in Health Services' was written for the Commission on Macroeconomics and Health [17]. These comprise three commentary / perspective journal pieces, four concept papers, and one literature review paper, using perspectives variously from Socio-legal studies, Economics and Management and Business Studies. The journals in which they were published ranged from public health to topic specialist, to social science publications. There was a lack of empirical studies in this area.

Smith et al's overview in the Lancet on trade in health-related services employs the GATS four models of service delivery as a classification for structuring their discussion [16]. In a thoroughly researched article they indicate that overall figures on the volume of such trade are vague, but that existing data indicates that India, along with the Philippines and Cuba, leads in exportation of medical-transcription services, tele-pathology, and tele-diagnostic services. The Apollo Group, the world's third largest private health provider, is given as an example with activities ranging from provision of regional telemedicine services from centres in South and Eastern Asia, to provision of back office services to US organisations for billing, documentation of clinical and administrative records, coding of medical processes and insurance claims processing. The authors lay out a considerable research agenda in concluding that the implementation and use of e-health raises issues about the recognition of credentials and licensure; legal liability and malpractice considerations, including provider insurance coverage; provider remuneration; patient privacy and confidentiality; the existence of enabling infrastructure and compatibility of standards in areas such as data, images, and medical records; infrastructure and future operating costs; and the quality and appropriateness and fragmentation of care.

Several of the articles identified dealt with the specific impacts on US jobs and salaries of outsourcing specific elements of health services to India [18, 19], and the issue of redress in cases of malpractice [20]. Stack and Downing, [19] for example, develop a framework to examine what types of healthcare and IT jobs in the USA are most and least at risk to offshoring. Their framework integrates key variables - the changing need for proximity, the regulatory environment, firm-level security, intellectual property rights concerns, and emphasises the need to understand how the decisions to offshore are shaped by broad socio-political forces. They then use this to examine the routing of between 4 & 10% of the US medical transcription industry to India, the potential offshoring of Ask-a-Nurse programmes by managed care companies, and offshore outsourcing of professional interpretation of radiographic imagery. Kshetri [21] suggests that the micro and macro level impacts of offshore outsourcing in general are far from clear and uses institutional theory as a lens through which to consider the issues raised by existing literature, arguing that this allows us to understand the 'macro level rules of the game' For Kshetri outsourcing activities can be viewed as arenas of power relations in which various players engage in 'institutional war' and such an approach can illuminate the mechanisms by which regulative rules, social rules, culturally supported habits influence assessment, selection and continuation of outsourcing projects [21].

Transnational consumption of services

India is part of a growing Asian market for medical or health 'tourism', attracting consumers from developing and industrialised countries [16, 22]. This topic has generated interest of late, twenty one papers published from 2004 to 2011 were found on this theme. The publications included medical and health related journals and social science journals, with a number of recent papers in *Global Social Policy*. Disciplinary perspectives include Management, Business Administration, Economics, Geography, Social Policy and Socio-legal studies. Overall the studies lack explicit theoretical orientation, and empirical studies were scarce until 2010. In all but one paper in this India-focused subset of a broader literature, medical tourism is considered from a health systems perspective. Connell [23] is the exception, taking a wider perspective from Geography to consider medical tourism as a niche market within tourism.

Several papers consider the implications of increasing medical tourism in India for national healthcare systems, either from the perspective of India or patients' country of origin. The potential accentuation of problems and inequalities within the Indian healthcare system as a result of medical tourism is raised in several commentary and review papers [22, 24, 25]. Vijaya [26] draws on a small number of interviews with nurses to highlight how the medical tourism market may take away resources that otherwise would benefit the public health system in India. Some limited consideration is given to the implications of medical tourism to healthcare systems in the US and UK [27, 28] and in these papers India is included as a major destination for those seeking care rather than the sole focus. In a further paper Cortez [29] provides a description of the legal system in India to assess what legal recourse is available to consumers if they fall victim to medical negligence when seeking care in another country.

Lunt and Carrera [30] provide a literature review on patient mobility from a European perspective but conclude that very little research is available in that area. However, our search does suggest that empirical social science research is increasing. Through thematic content analysis of marketing materials, Crooks et al. [31] study the messages and images used by companies to market India as a global health destination whilst Alsharif et al. [32] and Kangas [33] try to ascertain reasons for seeking care from the patients' perspective. Overcoming challenges in identifying patients seeking cross-border care, Alsharif et al. [32] uses a questionnaire survey administered by health professionals in four countries. In contrast Kangas [33] builds detailed qualitative case studies to understand the drivers for patients from a low income country, Yemen, to seek care in India.

Twelve papers focus on India as a major destination for 'reproductive tourism' or travel. Ikemotot [34] argues that although reproductive tourism is a subset of medical tourism it is distinct in a number of important ways. Assisted reproductive technologies (ART) rely on women's bodies or their reproductive labour, and if successful this results in the birth of a child, with implications that go beyond consumer's interests alone [34]. We found a number of commentaries highlighting ethical issues arising from the different levels of regulation that make India a leading destination for ART [35, 36]. Three papers from socio-legal studies [37, 38] and social policy [39] perspectives consider how surrogate mothers in India will be protected through the currently proposed ART Bill. The tone of discussions across papers is generally cautionary and looking for greater restriction of practice, although Crozier [40] takes a rather different stance arguing that the cross-border providers of ova and surrogacy may be better protected by increasing the alternative options available to them rather than restricting their participation in this market.

A recent critical narrative review of literature highlights that most papers on cross-border reproductive care are based on commentary rather than empirical research [41]. Empirical research in the area of reproductive tourism is challenging but, as with medical tourism more broadly, there are recent studies attempting to overcome this. Surveys of fertility patients (in Canada, US and Australia) to identify the scope and volume of cross-border fertility services confirm that India is a major site of oocyte donation [42, 43]. Pande provides the most substantial in-depth analysis with her study of surrogates in Gujarat [44, 45]. Drawing on oral histories of 42 surrogates and participant observation she draws on feminist and global production theory to introduce the concept of 'sexualised care work' to describe commercial surrogacy, highlighting the parallels that can be drawn with other care work and with the stigma ascribed to sex work.

Transnational movement of capital

Transnational movement of capital can take the form of loans or of foreign direct investment (FDI) for healthcare related provision. The very small literature in this area is dominated by perspectives from economics, and its size probably reflects difficulties in accessing data in this area. We identified six items published from 2001 onwards: two review articles, three articles and chapters using secondary data and two documents reporting on the findings from primary research. Together they shed light on the trends in this area over time. In the sole social policy piece Baru notes the role of the regional business groups and non-resident Indian doctors in promotion of the corporate sector in the southern cities of Hyderabad and Madras during the 1980s [46]. Smith indicates the evolution in the 1990s of new collaborations between Indian companies and multinational corporations such as Gleneagles based in Singapore, Royalton Medical Management based in Montreal, and Jardine Insurance from the UK [47]. In a subsequent paper Smith et al. report on the situation since 2000 in which India has allowed FDI in hospitals up to 100%, and they estimate that 90 FDI projects were approved (21 hospitals, 69 diagnostic centres) between

2000 and 2006, for a total US$53 million [16] However, Chanda [48] argues that increased FDI inflows have been impeded by regulatory and structural impediments in the Indian economy.

With regard to external loans, Chakravarthi [49] highlights the role of the International Finance Corporation (IFC), a member of the World Bank Group, also in promoting private sector involvement in health care in India by providing loans to Indian corporates for their expansion plans and to set up hospitals in smaller cities and towns within the country. This expansion activity ranged from investment in existing facilities, joint ventures with local partners, or managed care services which integrate the financing and the delivery of medical services and alliances to develop healthcare networks and chains.

Lefebvre [50] and Chakravarthi [49] also describes how Indian FDI export has grown, citing as an example the Apollo Group of Hospitals which has franchises in several countries in Asia and Africa and partnerships with hospitals in the Gulf States, Mauritius, Tanzania, Nigeria, UK, Sri Lanka, Bhutan, Pakistan, and Bangladesh, as well as a telemedicine centre in Kazakhstan. Several authors highlight the difficulties of procuring data in this topic area and of discrepancies in definitions. Outreville's international comparative study attempted to identify some of the determinants of foreign investment of the largest multinational companies (MNCs) and their relation with indicators of governance and stability, but encountered difficulty finding evidence on the relative importance of the largest MNCs operating in the healthcare sector [51]. Smith's review article argues that this research area still requires greater clarity on what constitutes 'FDI' in order then to assess the costs and benefits and to match them against country objectives [47]. Alternative definitions might include proportion ownership, foreign or local management systems, and investment on an affiliate or non-affiliate basis. In addition, Smith suggests that it may be important to distinguish between 'for profit' and 'not for profit' FDI, between sub-sectors such as hospitals, insurance or other areas such as medical education, and, especially in context such as India, between involvement of non-resident nationals and 'foreigners' involvement in FDI.

Chanda's work on the nature of foreign investment in hospitals in India stands out for its use of primary data [48,52]. Conducted for the World Health Organization, it utilises a survey of administrators and finance sections of 19 hospitals in 6 cities around the country, and semi-structured interviews with stakeholders ranging from health professionals to owners and managers of hospitals and nursing homes, health insurance companies, medical equipment suppliers and patients, civil society organizations and industry associations across three cities.

Transnational movement of healthcare workforce

Migration has historically been the main pathway of health services trade, following colonial and linguistic ties, and these patterns remain but are complicated by stepwise migration [53] and by intraregional and circular migration [54]. Such areas have received some attention from the social sciences, particularly in the last few years. India has been, and remains, a major source country for migratory health professionals and we identified 29 papers, chapters and reports published from 1989 onwards that made reference to transnational movement of Indian health professionals and met our other selection criteria, with over half of these published in the period 2007–9. This was the largest volume of work of any of the themes and reflected a disciplinary range across Sociology, Geography, Management studies, Education, Economics and Social Policy.

The majority of articles that we indentified are commentaries and overview papers drawing together existing data to review emerging trends and the implication for workforce planning and development [54, 55, 56, 57, 58, 59, 60,61, 62]. Chanda's chapter in Blouin et al. [63] reviews a range of regional and bilateral labour, trade and economic cooperation agreements and also national recruitment codes, considering their objectives, scope and degree of legal compulsion. Dovlo's overview paper [64] highlights the current interest in measuring and managing the migratory flow of health workers, in seeking reparations, payments or remittances, and in training substitute health workers. Kingma's extensive work on nurse migration [54, 60] brings together data from existing reports and studies to look at migration flows in the light of national nursing workforce imbalances, and to examine factors that encourage or inhibit nurse mobility, the potential benefits of circular migration, and nurse migration in the larger context of the global workforce. This synthesis is supplemented with interviews with nurse migrants on their motivations.

In order to understand migratory flows of physicians and nurses most authors draw on the data available from existing national professional governance systems and registers in receiving countries [16, 65, 66, 67, 68] and immigration systems [69]. Khadria's description of the emergence of nurse recruitment hubs in India draws on 'media reports' and on data on nurses taking the US Commission on Graduate

Foreign Nursing Schools examinations, to estimate the numbers of migrating nurses [70]. Only a few authors within our review had attempted comparative analyses for physician migration, using collated data from receiving countries [71] or a questionnaire to key informants in sending countries [65].

Nine of the twenty-nine papers and chapters report on empirical studies. These studies mainly focus on the motivations and experiences of the migrant doctor[69, 71] or nurse [53, 54, 55, 72, 73, 74, 75, 76]. A few papers deal with the cultural dimensions of migration of health workers. Robinson and Carey's series of repeat in-depth interviews with Indian doctors working in the UK NHS explore the multi-leveled nature of explanations for international migration [69] They suggest that while the movement of Indian doctors to the UK appears an economically driven phenomenon, migration also occurs because of "taken for granted" familial obligations rooted in the everyday cultural life and even the impressions generated by novels read in childhood. Three studies considered gender dimensions. Brush & Vasupuram [75] draw on existing literature to analyze the similarities and differences between skilled professional female migrants (nurses) and domestic workers (nannies and in-home caretakers) and how societal expectations, meanings, and values of care and 'women's work', together with social, cultural, economic and political processes, construct the female migrant care-giver experience. Nair and Percot [53] draw on multisite ethnographic studies in Delhi and the Gulf States to elaborate the nuanced nature of stepwise migration for female Malayali nurses from Kerala. Finally, in her 'global ethnography' about the lives in the USA of migrant nurses from the state of Kerala in India, sociologist Sheba George uses the experience of these nurses to explore how contradictory cultural values alter the balance of gender relations between couples in the migratory process, when women who migrate before men become the breadwinners in the family, and how upheaval occurs not only in the families and immigrant community but also in the sending community in India [77].

Three papers problematise other conventional wisdoms around migration of health workers and take an overtly global framing for their analytical stance. Wright et al's paper examines the historical antecedents to the current ethical debate around the 'brain drain' of health professionals [78]. S. Smith analyses the 'microcosm responses' adopted by one individual nation state, Australia, to the healthcare workforce shortage and argues for a coordinated global response with individual nation states being cognisant of the impact of their health policy, regulations and legislation on the 'global community' [79]. In one of the few papers to make explicit use of social science theory Raghuram, a geographer, offers an analysis of the dominant discourses around medical migration. In particular Raghuram focuses on the ways in which the brain drain discourse has privileged autonomous nation-states as the basis for, and the appropriate spatial unit for, thinking about the ethics of health care provision, and argues the case for a meso-level system whereby the benefits of skilled migration for destination countries would be calculated and an appropriate compensation paid to the countries from which migrants departed [80].

Production, consumption and trade in specific healthcare related commodities

The search highlighted that the Indian healthcare economy is associated with a number of new developments in specific marketable niches of practice, therapy, technologies or research and development, we term these collectively as 'healthcare related commodities'. Social science contributions to an understanding of this area could be found in relation to the organ trade [81, 82, 83, 84] stem cell research and therapy [85, 86, 87, 88, 89], non-allopathic practices [90, 91, 92, 93], biomedical clinical trials and biopharmaceutical research [94, 95, 96, 97, 98, 99,100] and clinical by-products [101]. The review identified 21 articles in total published from 1998 to 2010. The main disciplinary perspectives include Political Science, Anthropology, Geography, Sociology, and History.

A number of editorials and overview papers chart India as a growing market for the different healthcare related commodities. In the case of organ trade and stem cell research this is noted to be facilitated by levels of legislation or weak enforcement of legislation [[82, 83, 84, 85] and improvement to legislation is the focus of two editorials [81, 98]. The implications of growth in trade within a globalized market are also considered. In the case of outsourcing clinical trials in India concerns are raised around safety and the public health interests of the Indian population [96, 98] as well as the likely impact on development in outsourcing countries such as the US [94]. In the case of Ayurveda, concerns are raised that globalization and resulting standardisation have reduced Ayurveda from a system and overriding philosophy to itemised commodities to be traded on the global market [90, 91], and Hoyez [93] using ethnographic research in India and France, analyses the importance of place to the practice of yoga through the concept of 'therapeutic landscapes'.

A more in-depth understanding of globalization and the drivers and politics of these growing commodity markets, as well as potential implications for policy and for participants, is developed across the remaining papers. Salter & Salter [89] consider the political utility of bioethics in Human Embryonic Stem Cell (HESC) research. The paper analyses how the moral status of human embryos in particular creates the political need for bioethics to act as a vehicle for resolving cultural conflict and in further highlights the interests of countries such as India in adopting an approved bioethical position in order to legitimise its policy stance and global presences in HESC politics. With regard to stem cell research and regenerative medicine, Salter, a political scientist, demonstrates that in the absence of comprehensive resources countries such as India have to be selective about where they chose to intervene in order to maximise advantage. The explanatory concept of 'speculative states' is developed to understand India's intervention in the regenerative medicine knowledge production process [87, 88].

Patra & Sleeboom-Faulkner [86] examine how adult stem cell therapy is promoted in two leading centres in India. They use the concept of 'bionetworking' to examine how the recruitment of patients at local, national and global levels builds on the 'political economy of hope', from either patient or professional perspectives. In a similar vein, using a stakeholder analysis Muraleedharan [82] applies a politico-economic perspective to argue that the seemingly ineffective implementation of the Transplantation of Human Organs Act (1994) in India can be understood in the fact that all stakeholders can gain from the payment of organs and thus will continue to participate in the market. And in a highly original anthropological piece, Hodges [101] draws on cultural historical perspectives to understand the market in by-products of clinical encounters or 'medical garbage'. The analysis shows how different 'value chains' of people arise and that risk has a crucial mediating role in the value of medical garbage and work undertaken to extract its value. A willingness to take on risk through the handling of waste is compensated for by the opportunity to earn money in direct economic exchange.

Finally, a body of sociological and anthropological work argues for a more critical understanding of participation in clinical trials [95, 97, 99, 100]. For Prasad the rights and values of the population participating in trials should be understood in a transnational market in which India is a competitor [97]. Cooper [95] applies the concept of 'experimental labour' to account for some of the groundwork through which value is created in biomedical economies. Sunder Rajan focuses on ethics and argues that the harmonization of ethics goes with the harmonization of property regimes globally and as a result provide global capital that can turn "even healthy Indian populations into experimental subjects, who are both merely risked and free to choose to be so" 74: [100].

New global governance issues and structures

Increased transnational activities have brought new transnational regulations and policies, some of which impact upon the healthcare arena in new ways. The search identified fourteen relevant articles, all published from 2000 onwards. The majority focused on the consequences for India of international agreements administered by the World Trade Organization (WTO) such as the agreement on Trade Related Aspects of Intellectual Property Rights (TRIPS) which sets minimum standards for intellectual property regulation and the General Agreement on Trade in Services (GATS). Disciplinary perspectives included Economics, Management, Law and Political Science. Sources ranged from *the Lancet* to social science journals and World Bank Reports.

One paper considered broader global health governance. From a socio-legal perspective, Gostin [102] provides a critical analysis of current global health architecture and need for change in order to meet basic health needs and to ensure greater country ownership and autonomy within health. With regards to TRIPS, common themes were the analysis of implications of new patent laws on access to medicines, particularly antiretrovirals [103, 104] and of the emerging opportunities and challenges for pharmaceutical industries [105, 106, 107, 108, 109, 110]. Papers consider how India has responded to TRIPS, how patent policies have been amended and implemented within India, and to what extent innovation and research and development has been stimulated following increased patent protection. There were two microeconomic analyses. Fink [107] calibrates a theoretical model with data from the pharmaceutical industry in India to simulate the possible impact of stronger patent policies on transnational corporations and in turn on market and welfare impacts, and Chadha [105] draws on product cycle and neo-technology theories of trade to analyse export performance of 131 Indian pharmaceutical companies.

Papers in this theme highlight the potential tensions between adherence to international agreements and achievement of national health goals, although India is given only limited focus within broader discussion.

With regards to GATS, overviews by Chanda and Smith referred to earlier consider the implications of trade in health services, both positive and negative, for key health policy goals including equity, efficiency, quality and access [15,16]. Both point to the need for improved data collection and studies in order to answer issues raised by increased trade in health. Whittaker [36], in a case study of reproductive tourism in Asia, argues that GATS and current global regulation are not sufficient to address problems of exploitation and how to balance reproductive autonomy of individuals with the protection of rights of others.

Bloche [111] carefully analyses the laws and WTO's response to argue that the potential threat to health and prioritisation of the protection of health within countries is misunderstood, and that there are social and political reasons why health protection has actually become more central to WTO even if not explicitly stated. In an exploratory discussion Chanda [52] is also more positive suggesting that GATS has further potential as a tool to ensure that the movement of health professionals is temporary rather than permanent and as a result be beneficial to host and source countries.

Transnational diffusion of ideas and knowledge

The final theme concerns the transnational diffusion of ideas and knowledge, including policy ideas, political ideologies and discourses around health and healthcare in the context of India. In total there were nineteen articles (including three book chapters and two conference papers) and it is in this theme that India based scholars are particularly prominent. There was a strong emphasis on analysis of the influence of neoliberalism on health and healthcare delivery systems.

Published from 1998 to 2010, disciplines represented were primarily Social Policy, Sociology and Area and Development Studies, with smaller contributions from Management, Social Anthropology and Political Science. Journals of publication related to Health and Development, Gender Studies and Global Health, and Business Administration.

Two papers using contrasting quantitative and qualitative approaches attempt to study the impact of globalisation and economic reforms on the health of workers and reach similar conclusions that work and income are less reliable at the same time as out-of-pocket expenditure is increasing due to the growth in private health services [112, 113]. Analysis from Social Policy shows that increasing privatisation of healthcare led to reductions in public expenditure on health and that this disproportionately impacts on vulnerable groups [46, 113, 114, 115, 116]. Rao [116] illustrates that whilst globalisation has led to international policy such as the Programme of Action promoting reproductive rights and health resulting from the International Conference on Population and Development (ICPD), diffusion of neoliberal ideas can at the same time erode the possibility of such gains, through the collapse of the India public health system. Two further papers consider the influence on neo-liberalism on autonomy and independence in decision making. Qadeer [117] argues that globalisation and uniformity of programmes means that it is very difficult for national governments in countries such as India to take autonomous decisions concerning reproductive health programmes, and Ramachandran [118] provides a critique of the reliance of NGOs on donors for project grants and how this has distanced them from their activist roots.

Two multi-country studies from political scientists examine the adoption of transnational policy ideas [119, 120]. Through a systematic analysis of country scoping reports Holden [119] examines the export strategy of Public-Private Partnerships (PPP) from the UK. Shiffman [120] draws on qualitative case studies of five countries, including India, to understand what factors determine the political priority given to maternal mortality.

Donor and global discourses around HIV/AIDS and reproductive health forms another sub group within this theme and draw analysis from Sociology, Anthropology, Social Policy and Politics and International Studies. Sengupta [62] provides an analysis of the World Bank reproductive health programme in India, showing that the programme incorporates language used by feminist and radical groups who have opposed population control programmes in India but that the programme has not substantively changed. Finn & Sarangi [121] critically appraise the appropriateness of neoliberal, western-centric discourses on the attainment of health related quality of life and individual risk behaviours in the context of HIV in India. Using the UK's Department of International Development and Norwegian's Agency for Development Co-operation as case examples, Jones [122] tries to deconstruct donor policy to understand the predominate focus on prevention rather than treatment of HIV/AIDS. Misra's ethnographic study examines how globally established scientific material is delivered locally through transnational exports and in turn how local translation feeds back to the global level [123].

The transnational reach of pharmaceutical companies is the focus of three studies within this theme. The first by Bower & Sulej focuses specifically on the growth of Indian pharmaceutical companies into western markets, examining the explanatory concepts of absorptive capacity, social capital and strategic alliances in cross-national contexts [124]. The second by Haakonsson uses global value chain analysis to understand the social and political pressures involved in production and trade to examine opportunities for upgrading pharmaceutical industry for which changes in Indian market provide important opportunities [125]. Ecks [126] introduces the concept of 'near-liberalism' to emphasise that elements of neo-liberalism, freedom, emancipation and autonomy are always held out as future goals. The anthropological case study of marketing of antidepressants shows how a company's stated global corporate citizenship can be easily adapted in practice.

DISCUSSION

Using the lens of India, this literature review attempts to capture the 'state of the art' of social science literature concerning the evolving global healthcare landscape. Inevitably, there are several caveats. Such a far reaching topic as this does not lend itself to easy classifications. Despite the multiple search steps we cannot claim to have identified all the social science scholarship in this field, only that we have made a fairly comprehensive and systematic attempt. It is possible that some important work, most likely book chapters, will have eluded our methods of searching. The restriction to English language is an additional constraint. What became clear through our screening process is that the boundaries of what we determine to be 'social science' in the health sphere are porous. Ideas from the social sciences become mainstreamed (not least 'globalisation' itself). Social policy perspectives and social science research methods are taken up and employed by health professionals. We tried to make consistent decisions on where to draw our boundary for this review process, others might redraw it a little differently.

The specific focus on work related to a middle income country such as India both illuminates and restricts. A country focus did permit a more explicit search for work on the interface between the macro and the micro, the global and the local, than would have been feasible otherwise (our step 5). And it illuminates because it helps us to think through the inward and outward processes of transnational activity and their impact upon a specific locality. It restricts because it implicitly retains an orientation to the national state and because it may exclude seminal theory and empirical work that did not focus on that particular locality.

While it is impossible here to elaborate a research agenda for all the themes or all the social science disciplines here, several comments can usefully be made about what is still to be done and the research challenges of doing it. Firstly, the inadequacy of global data and the difficulties with definitions and data collection methods on all dimensions of health-related trade has been highlighted and this constraint was also apparent in much of the public health and health policy literature that we scanned and discarded during the selection process. This situation inhibits a proper understanding of global trends and comparative analysis, and requires attention by researchers and at a policy level. Secondly, although we found some evidence of cross-disciplinarity with reference made to other literatures from sister disciplines, the field is ripe for more cross-fertilisation. In particular we would endorse the call made by several key writers for empirical studies that integrate social and economic perspectives more explicitly. Third, the value of historical and contextualising analyses that unsettle the prevailing policy discourses has been demonstrated by a small number of scholars, but such work is scarce in spite of its clear relevance to India's relationships with global forces.

Fourth, there is scope for many more empirical studies, and for greater testing and development of explanatory theory. In most thematic areas we found the majority of the literature to be overviews or commentaries. Quite a number of these did not move beyond the descriptive, although others did make careful and thorough critique of the available data. We did find interesting empirical work exploring the social processes inherent in the development and adoption of new technologies and those examine new transnational ways of working. It is perhaps the work grouped under the theme of 'specific healthcare related commodities' that offers some of the most illuminating social science work on health systems and globalisation within the body of literature that we have reviewed. Whilst the overview papers and commentaries point up potential ethical and governance problems, the body of anthropological and biopolitical research offers a rather deeper understanding of how actors at local, national and global levels can, for example, take advantage of unequal social and regulatory contexts to further stem cell research, and of the enactment of power and policy. Ethnographic studies of specific 'products' and of the production and reproduction of therapeutic landscapes are challenging and resource intensive but ultimately rewarding.

Underlying many of the preceding points is the broad question of how best social science researchers can connect the broad discussions of globalization with the local realities in health systems. Across the literature that we reviewed we saw very little comprehensive analysis of the kind that Holden and others have argued for elsewhere [127] – analyses that use multilevel data relating to the level of the firm or organisation, to the national level, to regional processes and institutions and to international organisations and agreements, and the interactions between these. Such an approach would seem to be a promising avenue for the future. The extended case study method [128], represented by only one study in this review [74], would also seem to have considerable potential for the study of globalisation and health topics. The ambition of the approach - as Burawoy puts it is 'to extract the general from the unique, to move from the 'micro' to the 'macro', and to connect the present to the past in anticipation of the future' 5:[128]. This may be what will be required to link large-scale processes, transnational actors, and the fine grain of everyday praxis.

This is a young and expanding field of study, with over half of the identified work published from 2007 onwards. Less than a fifth of the items included were reports of empirical studies. Transnational consumption of health services, the transnational healthcare workforce, trade in specific health-related commodities, and transnational diffusion of ideas have all received attention from social scientists in work related to India. Global health governance, transnational delivery of health services and the transnational movement of capital have received less attention. No one discipline or journal dominates the field - the diversity across specialist, multidisciplinary, social science and health journals is striking. *Social Science and Medicine* was the biggest single contributor, but with only six articles, followed closely by *Global Social Policy* and the Indian journal *The Economic and Political Weekly*. The body of work that we have reviewed on India illuminates many ways in which local healthcare systems are ever more interconnected and interdependent with each other, and how intricately they have become woven into the economic, political, social, cultural and technological dimensions of globalisation.

ACKNOWLEDGEMENTS

The Economic and Social Research Council (UK) for funding support to Dr Bisht's fellowship; RES-075-25-0015 India's Challenge in a Globalizing Healthcare Economy: Social Science Directions. Dr Indira Charavarthy and Dr Altaf Virani for assistance in hand and web searching in India

COMPETING INTERESTS

The author(s) declare that they have no competing interests.

AUTHORS' CONTRIBUTION

SFM and EP provided guidance on the framework and direction of the literature review. RB conducted the search, reviewed articles, prepared drafts of tables and drafted the methods section. EP reviewed articles contributed to the synthesis, wrote and redrafted manuscripts. SFM reviewed articles, contributed to the synthesis, and provided substantive inputs on drafts of the manuscript. All authors were involved in reading drafts of the manuscript and providing comments and suggestions for the paper. All authors read and approved the final manuscript.

REFERENCES

1. Walby S: Globalization & Inequalities: Complexity and Contested Modernities. 2009, London, Thousand Oaks, New Delhi, Singapore: Sage

2. World Health Organization: World Health Report, Health systems: improving performance. 2000, Geneva: WHO

3. Vaguet A (Ed): Indian Health Landscapes Under Globalisation. 2009, New Delhi: MANOHAR: Centre de Sciences Humaines

4. Ram-Persaud C: India's middle class failure. Prospect Magazine. 2007, 138-http://www.prospectmagazine.co.uk/article_details.php?id=9776,

5. Pricewaterhouse Coopers: Emerging market report: healthcare in India. 2007

6. Grace C: Update on China and India and Access to Medicines. A Briefing Paper. 2005

7. UNDP: Human Development Report 2011 Sustainability and Equity: A Better Future for All. 2011, New York: UNDP

8. UNDP: Human Development Report 2010 The Real Wealth of Nations. Pathways to Human Development. 2010, New York: UNDP

9. Dilip T: Extent of Inequity in Access to Health Care Services in India. Review of Health Care in India. Edited by: Gangolli L, Duggal R, Shukla A. 2005, Mumbai: Centre for Enquiry into Health and Allied Themes, 247-268.

10. Krishnan TN: Access to Health and the Burden of Treatment: An international Comparison. Disinvesting in Health: The World Bank's Prescription for Health. 1999, New Delhi: Sage Publications

11. Sen G, Iyer A, Asha George A: Structural Reforms and Health Equity-Acomparison of NSS Surveys, 1986–87 and 1995–96. Econ Pol Wkly. 2002, 342-1352.

12. Banerji D: Simplistic Approach to Health Policy Analysis: World Bank Team on Indian Health Sector. Econ Pol Wkly. 1993, 28 (24): 1207-1210.

13. Rao S: Financing and Delivery of Health Services in India. 2005, New Delhi: Ministry of Health and Family Welfare, Government of India

14. Gilson L, Raphaely N: The terrain of health policy analysis in low and middle income countries: a review of published literature 1994–2007. Health Policy Plan. 2008, 23 (5): 294-307.

15. Chanda R: Trade in health services. Bull World Health Organ. 2002, 80 (2): 158-163.

16. Smith RD, Chanda R, Tangcharoensathien V: Trade in health-related services. Lancet. 2009, 373 (9663): 593-601.

17. Chanda R: Trade in health services. 2001, Prepared for the Working Group on Health and International Economy of the Commission on Macroeconomics and Health, ICRIER Working paper No 70 (revised)

18. McLean TR: Will India set the price for teleradiology?. The International Journal of Medical Robotics and Computer Assisted Surgery. 2009, 5 (2): 178-183.

19. Stack M, Downing R: Another look at offshoring: Which jobs are at risk and why?. Bus Horiz. 2005, 48 (6): 513-523.

20. Zawadski P: International outsourcing plus inexpensive, quality healthcare: binding arbitration makes this telemedical dream a reality. Health Matrix. 2008, 18 (1): 137-179.

21. Kshetri N: Institutional factors affecting offshore business process and information technology outsourcing. J Int Manag. 2007, 13 (1): 38-56.

22. Blouin C: Trade in health services: Can it improve access to health care for poor people?. Global Social Policy. 2010, 10 (3): 293-295.

23. Connell J: Medical tourism: Sea, sun, sand and ... surgery. Tour Manage. 2006, 27 (6): 1093-1100.

24. Godwin SK: Medical Tourism: Subsidising the Rich. Econ Pol Wkly. 2004, 3981: 83-

25. Reddy S, Qadeer I: Medical Tourism in India: Progress or Predicament?. Econ Pol Wkly. 2010, 45 (20): 69-75.

26. Vijaya RM: Medical Tourism: Revenue Generation or International Transfer of Healthcare Problems?. Journal of Economic Issues. 2010, 44 (1): 53-70.

27. Cortez N: Patients without borders: The emerging global market for patients and the evolution of modern health care. Indiana Law Journal. 2008, 83 (1): 71-132.

28. Horowitz MD, Rosensweig JA, Jones CA: Medical tourism: Globalization of the healthcare marketplace. Medscape General Medicine. 2007, 9 (4): 33-

29. Cortez N: Recalibrating the Legal Risks of Cross-Border Health Care. Yale J Health Policy Law Ethics. 2010, 1:

30. Lunt N, Carrera P: Medical tourism: Assessing the evidence on treatment abroad. Maturitas. 2010, 66 (1): 27-32.

31. Crooks VA, et al: Promoting medical tourism to India: Messages, images, and the marketing of international patient travel. Soc Sci Med. 2011, 72 (5): 726-732.

32. Alsharif MJ, Labonte R, Zuxun L: Patients beyond borders: A study of medical tourists in four countries. Global Social Policy. 2010, 10 (3): 315-335.

33. Kangas B: The burden of pursuing treatment abroad: Three stories of medical travellers from Yemen. Global Social Policy. 2010, 10 (3): 306-314.

34. Ikemotot LC: Reproductive Tourism: Equality Concerns in the Global Market for Fertility Services. Law and Inequality. 2009, 27:

35. Saronjini NB, Das D: ARTs: Voices from Progressive Movements. Making Babies. Edited by: Srinivasan S. 2010, New Delhi Zubaan, 22-43.

36. Whittaker A: Challenges of medical travel to global regulation: A case study of reproductive travel in Asia. Global Social Policy. 2010, 10 (3): 396-415.

37. Rimm J: Booming baby business: regulating commercial surrogacy in India. University of Pennsylvania Journal of Comparative Corporate Law. 1429-1462. 2008–2009

38. Smerdon UR: Crossing bodies, crossing borders: international surrogacy between the United States and India. Cumberland Law Review. 2009, 39 (1): 15-84.

39. Palattiyil G, et al: Globalization and cross-border reproductive services: Ethical implications of surrogacy in India for social work. Int Soc Work. 2010, 53 (5): 686-700.

40. Crozier G: Protecting cross-border providers of ova and surrogacy services?. Global Social Policy. 2010, 10 (3): 299-303.

41. Hudson N, et al: Cross-border reproductive care: a review of the literature. Reproductive BioMedicine Online. 2011, 22 (7): 673-685.

42. Blyth E: Fertility patients' experiences of cross-border reproductive care. Fertil Steril. 2010, 94 (1): e11-e15.

43. Hughes EG, DeJean D: Cross-border fertility services in North America: a survey of Canadian and American providers. Fertil Steril. 2010, 94 (1): e16-e19.

44. Pande A: Not an 'Angel', not a 'Whore': Surrogates as 'Dirty' Workers in India. Indian Journal of Gender Studies. 2009, 16 (2): 141-173.

45. Pande A: Commercial Surrogacy in India: Manufacturing a Perfect Mother-Worker. Signs: Journal of Women in Culture and Society. 2010, 35 (4): 969-992.

46. Baru RV: Health Sector Reforms and Structural Adjustment: a State-level analysis. Public health and the poverty of reforms: the South Asia predicament. Edited by: Qadeer I, Sen K, Nayar KR. 2001, New Delhi: Sage Publications, 211-234.

47. Smith RD: Foreign direct investment and trade in health services: A review of the literature. Soc Sci Med. 2004, 59 (11): 2313-2323.

48. Chanda R: Constraints to foreign direct investment in Indian Hospitals. Journal of International Commerce, Economics and Policy. 2010, 1 (1): 121-143.

49. Chakravarthi I: Corporate Presence in the Healthcare Sector in India. Social Medicine. 2010, 5 (4): 192-204.

50. Lefebvre B: 'Bringing World-class health care to India': The Rise of Corporate Hospitals. Indian Health Landscapes Under Globalisation. Edited by: Vaguet A. 2009, New Delhi: MANOHAR Centre de Sciences Humaines, 83-99.

51. Outreville JF: Foreign direct investment in the health care sector and most-favoured locations in developing countries. Eur J Health Econ. 2006, 8 (4): 305-312.

52. Chanda R: Foreign Investment in Hospitals in India: Status and Implications. 2007, India

53. Nair S, Percot M: Transcending Boundaries: Indian nurses in national and international migration. 2007, Available at: http://www.cwds.org

54. Kingma M: Nurses on the Move: A Global Overview. Health Serv Res. 2007, 42 (3p2): 1281-1298.

55. Aiken LH: Nurse Labor Market Dynamics Are Key to Global Nurse Sufficiency. Health Serv Res. 2007, 42 (3 Pt 2): 1299-1320.

56. Batata AS: International nurse recruitment and NHS vacancies: a cross-sectional analysis. Glob Heal. 2002, 1: 7-7.

57. Brush BL: Global Nurse Migration Today. J Nurs Scholarsh. 2008, 40 (1): 20-25.

58. Buchan J: International Recruitment of Nurses: Policy and Practice in the United Kingdom. Health Serv Res. 2007, 42 (3p2): 1321-1335.

59. Hallock JA, McKinley DW, Boulet JR: Migration of doctors for undergraduate medical education. Medical Teacher. 2007, 29 (2–3): 98-105.

60. Kingma M: Nurses on the Move. Migration and the Global Health Care Economy. 2006, Ithaca & London: Cornell University Press

61. Saravia NG, Miranda JF: Plumbing the brain drain. Bull World Health Organ. 2004, 82 (8): 608-615

62. Sengupta A: World Bank's Reproductive and Child Health Approach: Old Wine in New Bottles. Reproductive Health in India's Primary Health Care. Edited by: Rao M. 1998, New Delhi: Centre of Social Medicine and Community Health, Jawaharlal Nehru University, 31-40.

63. Chanda R: Trade Policy and Health Professionals' Movement. Trade and Health: Seeking Common Ground. Edited by: Blouin

C, Heymann J, Drager N. 2007, Ottawa, Ontario: McGill University Press, 93-129.

64. Dovlo D: Taking More Than a Fair Share? The Migration of Health Professionals from Poor to Rich Countries. PLoS Med. 2005, 2 (5): e109-

65. Astor A, Akhtar T, Matallana MA, et al: Physician migration: Views from professionals in Colombia, Nigeria, India, Pakistan and the Philippines. Soc Sci Med. 2005, 61 (12): 2492-2500.

66. Blythe J, Baumann A: Internationally educated nurses: profiling workforce diversity. Int Nurs Rev. 2009, 56 (2): 191-197.

67. Humphries N, Brugha R, McGee H: Overseas nurse recruitment: Ireland as an illustration of the dynamic nature of nurse migration. Health Policy. 2008, 87 (2): 264-272.

68. Rockett IR, Putnam SL: Physician-nurse migration to the United States: regional and health status origins in relation to legislation and policy. International Migration. 1989, 27 (3): 389-409.

69. Robinson V, Carey M: Peopling Skilled International Migration: Indian Doctors in the UK. Int Migr. 2000, 38 (1): 89-108.

70. Khadria B: International Nurse Recruitment in India. Health Services Research. 2007, 42 (3): 1429-1436.

71. Kangasniemi M, Winters LA, Commander S: Is the medical brain drain beneficial? Evidence from overseas doctors in the UK. Soc Sci Med. 2007, 65 (5): 915-923.

72. Hawthorne L: Qualifications recognition reform for skilled migrants in Australia: Applying competency-based assessment to overseas-qualified nurses. Int Migr. 2002, 40 (6): 55-92.

73. Alonso-Garbayo Á, Maben J: Internationally recruited nurses from India and the Philippines in the United Kingdom: the decision to emigrate. Hum Resour Heal. 2009, 7 (1): 37-

74. George S: "Dirty Nurses" and "Men who Play". Gender and class in transnational migration. Global Ethnography: Forces, Connections, and Imaginations in a Postmodern World. Edited by: Michael B, Joseph A, Blum M, Sheba G, Zsuszsa G, Teresa G,

Lynne H, Maren K, Steven H, Lopez Z, Riain SO, Millie T. 2000, University of California Press, 144-174.

75. Brush BL, Vasupuram R: Nurses, nannies and caring work: importation, visibility and marketability. Nurs Inq. 2006, 13 (3): 181-185.

76. Centre for Trade and Development (CENTAD) in collaboration with WHO India Office: Migration of Health care professionals from India – a case study of nurses. 2009, Delhi: CENTAD with WHO India office

77. George SM: When Women Come First. Gender and Class in Transnational Migration. 2005, Berkeley: University of California Press, 1-296.

78. Wright D, Flis N, Gupta M: The "Brain Drain" of physicians: historical antecedents to an ethical debate, c. 1960–79. Philos Ethics Humanit Med. 2008, 3: 24-24.

79. Smith S: The global workforce shortages and the migration of medical professions: the Australian policy response. Australia and New Zealand Health Policy. 2008, 5 (1): 7-10.1186/1743-8462-5-7.

80. Raghuram P: Caring about "brain drain" migration in a postcolonial world. Geoforum. 2009, 40 (1): 25-33.

81. Daar AS: Paid organ donation – the grey basket concept. J Med Ethics. 1998, 24 (6): 365-368.

82. Muraleedharan VR, Jan S, Ram Prasad S: The trade in human organs in Tamil Nadu: the anatomy of regulatory failure. Health Econ Policy Law. 2006, 1: 41-57.

83. Roberts ED: When the Storehouse is Empty, Unconscionable Contracts Abound: Why Transplant Tourism Should Not Be Ignored. Howard Law Journal. 2009, 52 (3): 747-789.

84. Turner L: 'Medical tourism' initiatives should exclude commercial organ transplantation. JRSM. 2008, 101 (8): 391-394. 10.1258/jrsm.2008.080058.

85. Mason C, Manzotti E: Stem cell nations working together for a stem cell world. Regen Med. 2010, 5 (1): 1-4. 10.2217/rme.09.85.

86. Patra PK, Sleeboom-Faulkner M: Bionetworking: experimental stem cell therapy and patient recruitment in India. Anthropol Med. 2009, 16 (2): 147-163.

87. Salter B: Governing stem cell science in China and India: emerging economies and the global politics of innovation. New Genetics and Society. 2008, 27 (2): 145-159.

88. Salter B: State Strategies and the Geopolitics of the Global Knowledge Economy: China, India and the Case of Regenerative Medicine. Geopolitics. 2009, 14 (1): 47-78.

89. Salter B, Salter C: Bioethics and the Global Moral Economy: The Cultural Politics of Human Embryonic Stem Cell Science. Science Technology Human Values. 2007, 32 (5): 554-581.

90. Banerjee M: Public Policy and Ayurveda. Econ Pol Wkly. 2002, 37 (12):

91. Banerjee M: Local knowledge for world market. Globalising Ayurveda. Econ Pol Wkly. 2004, 89-93.

92. Hoyez A: Health, Societies and Identities in the Globalization Era: The example of yoga therapy. Indian Health Landscapes under Globalization. Edited by: Vaguet A. 2009, New Delhi: MANOHAR; Centre de Sciences Humaines

93. Hoyez A: The 'world of yoga': The production and reproduction of therapeutic landscapes. Soc Sci Med. 2007, 65 (1): 112-124.

94. Cockburn IM, Slaughter MJ: The global location of biopharmaceutical knowledge activity: New findings. New questions. Innovation Policy and the Economy. 2009, 10: 129-157.

95. Cooper M: Experimental Labour—Offshoring Clinical Trials to China. East Asian Science, Technology and Society: an International Journal. 2008, 2 (1): 73-92.

96. Dandona L: Clinical Trials in India: Balancing economic opportunity with the public health context. Natl Med J India. 2006, 19 (2): 55-57.

97. Prasad A: Capitalizing Disease: Biopolitics of Drug Trials in India. Theory, Culture & Society. 2009, 26 (5): 1-29.

98. Sen Gupta A: Fatal Trials: Clinical Trials are killing people (editorial). Indian J Med Ethics. 2009, 6 (3): 118-119.

99. Sunder Rajan K: Biocapital: the constitution of postgenomic life second. 2006, Durham and London: Duke University Press

100. Sunder Rajan K: The experimental machinery of global clinical trials: case studies from India. Asian biotech: ethics and communities of fate. Edited by: Ong A, Chen NN. 2010, Durham and London: Duke University Press

101. Hodges S: Chennai's Biotrash Chronicles: Chasing the Neo-Liberal Syringe. 2008, Warwick: University of Warwick

102. Gostin LO: Meeting basic survival needs of the world's least healthy people: Toward a framework convention on global health. Georgetown Law Journal. 2008, 96 (2): 331-392.

103. Arora GK, Gumber A: Globalisation and healthcare financing in India: some emerging issues. Public Finance & Management. 2005, 5 (4): 567-596.

104. Lanoszka A: The Global Politics of Intellectual Property Rights and Pharmaceutical Drug Policies in Developing Countries. International Political Science Review. 2003, 24 (2): 181-197.

105. Chadha A: Product Cycles, Innovation, and Exports: A Study of Indian Pharmaceuticals. World Dev. 2009, 37 (9): 1478-1483.

106. Chaudhuri S: R&D for development of new drugs for neglected diseases in India. Int J Technol Glob. 2010, 5 (1/2): 61-75.

107. Fink C: How Stronger Patent Protection in India Might Affect the Behavior of Transnational Pharmaceutical Industries. 2000, SSRN eLibrary, Available at: http://papers.ssrn.com/sol3/papers.cfm?abstract_id=630724

108. Guennif S, Chaisse J: Present stakes around patent political economy: Legal and economic lessons from the pharmaceutical

patent rights in India. Asian Journal of WTO and International Health Law and Policy. 2007, 2 (1): 65-98.

109. Hein W, Kohlmorgen L: Global Health Governance: Conflicts on Global Social Rights. Global Social Policy. 2008, 8 (1): 80-108.

110. Mueller JM: The Tiger Awakens: The Tumultuous Transformation of India's Patent System and the Rise of Indian Pharmaceutical Innovation. SSRN eLibrary. Available at: http://papers.ssrn.com/sol3/papers.cfm?abstract_id=923538&rec=1&srcabs;=630724

111. Bloche MG: WTO Deference to National Health Policy: Toward an Interpretive Principle. J Int Economic Law. 2002, 5 (4): 825-848.

112. Kumar S, Acharya SS: Workers' Health and the process of Exploitation in the era of liberalization: A case of Brassware Industry of Moradabad. Journal of Health and Development. 2006, 2 (1 & 2): 99-118.

113. Thankappan KR: Some health implications of globalization in Kerala, India. Bull World Health Organ. 2001, 79 (9): 892-893.**PubMedPubMed Central**

114. Qadeer I: Ethics and Medical Care in a Globalising World: Some Reflections. Restructuring Health Services: Changing Contexts and Comparative Perspectives. 2003, London: Zed Publications, 53-64.

115. Karlekar M: A case study from health sector. Women and Economic Reform in India. 1998, 1-23. Available at: http://www.cwds.org

116. Rao M: Cairo Door Ast? Population Policies and their context in India after ICPD. Indian Journal of Gender Studies. 2006, 13 (2): 247-274.

117. Qadeer I: Population Control in the era of neo-liberalism. Journal of Health and Development. 2005, 1 (4): 31-38.

118. Ramachandran V: NGOs in the Time of Globalisation. Reproductive Health in India's Primary Health Care. Edited by: Rao M. 1998, New Delhi: Centre of Social Medicine and Community Health, Jawaharlal Nehru University, 31-40.

119. Holden C: Exporting public-private partnerships in healthcare: export strategy and policy transfer. Policy Studies. 2009, 30 (3): 313-332.

120. Shiffman J: Generating Political Priority for Maternal Mortality Reduction in 5 Developing Countries. Am J Public Health. 2007, 97 (5): 796-803.

121. Finn M, Sarangi S: Quality of life as a mode of governance: NGO talk of HIV 'positive' health in India. Soc Sci Med. 2008, 66 (7): 1568-1578.

122. Jones PS: When 'development' devastates: donor discourses, access to HIV/AIDS treatment in Africa and rethinking the landscape of development. Third World Quarterly. 2004, 25 (2): 385-

123. Misra K: Politico-moral transactions in Indian AIDS service: confidentiality, rights and new modalities of governance. Anthropology Quarterly. 79 (1): 33-74.

124. Bower DJ, Sulej JC: The Indian Challenge: The Evolution of a Successful New Global Strategy in the Pharmaceutical Industry. Technology Analysis & Strategic Management. 2007, 19 (5): 611-624.

125. Haakonsson SJ: 'Learning by importing in global value chains: upgrading and South-South strategies in the Ugandan pharmaceutical industry. Dev South Afr. 2009, 26 (3): 499-516.

126. Ecks S: Near-liberalism: global corporate citizenship and pharmaceutical marketing in India. Asian biotech: ethics and communities of fate. Edited by: Ong A Chen NN. Durham and London: Duke University Press

127. Holden C: Investigating International Health Markets: Methodological Problems and Challenges. Social Policy and Society. 2005, 4 (04): 427-

128. Burawoy M: The Extended Case Method. Sociological Theory. 1998, 16 (1): 4-33.

7

CHAPTER

"You're Dealing with an Emotionally Charged Individual...": An Industry Perspective on the Challenges Posed by Medical Tourists' Informal Caregiver-Companions

Victoria Casey[1*], Valorie A Crooks[1], Jeremy Snyder[2] and Leigh Turner[3]

[1]Department of Geography, Simon Fraser University, 8888 University Drive,Burnaby, Canada.

[2]Department of Health Sciences, Simon Fraser University,8888 University Drive, Burnaby, Canada.

[3]Center for Bioethics, School ofPublic Health, and College of Pharmacy, University of Minnesota,Minneapolis, USA.

ABSTRACT

Background

Patients engage in medical tourism when they privately obtain a medical care abroad. Previous research shows that many medical tourists travel abroad with friends and family members who provide support and assistance. Meanwhile, very little is known about this important stakeholder group, referred to here as caregiver-companions. In this article we examine the challenges that can be posed by caregiver-companions and the overall practice of informal caregiving in medical tourism from an industry perspective. Specifically, we report on the findings of interviews conducted with international patient coordinators (IPCs) who work at destination facilities. IPCs come into regular contact with caregiver-companions in their professional positions and thus are ideally suited to comment on trends they have observed among this stakeholder group as well as the challenges they can pose to medical tourists, health workers, and facilities.

Methods

We conducted 20 semi-structured interviews with 21 IPCs from 16 different facilities across nine countries. Topics probed in the interviews included caregiver-companion roles, IPCs' and others' interaction with caregiver-companions, and potential health and safety risks posed to medical tourists and caregiver-companions. Thematic analysis of the verbatim transcripts was employed.

Results

Although most participants encouraged medical tourists to travel with a caregiver-companion, many challenges associated with caregiver-companions were identified. Three themes best characterize the challenges that emerged: (1) caregiver-companions require time, attention and resources; (2) caregiver-companions can disrupt the provision of quality care; and (3) caregiver-companions can be exposed to risks. IPCs pointed out that caregiver-companions may, for example, have a negative impact on the patient through cost of accompaniment or inadequate care provision. Caregiver-companions may also create unanticipated or extra work for IPCs, as additional clients and by ignoring established organizational rules, routines, and expectations. Furthermore, caregiver-companions may be susceptible to stresses and health and safety risks, which would further deteriorate their own abilities to offer the patient quality care.

Conclusions

Although caregiver-companions can pose challenges to medical tourists, health workers, and medical tourism facilities, they can also assist in enhancing best care and offering meaningful support to medical tourists. If caregiver-companions are open to collaboration with IPCs, and particularly in the form of information sharing, then their experience abroad can be safer and less stressful for themselves and, by extension, for the accompanied patients and facility staff.

Keywords: Medical tourism Informal caregiving International patient coordinators Patient International health care

BACKGROUND

Medical tourism involves patients travelling abroad to obtain non-emergency private medical care outside of established cross-boarder care arrangements [1, 2]. Because this medical care is private, usually the patient covers the costs of treatment and travel out of pocket [1, 3]. Individuals engage in medical tourism for a variety of reasons, including the desire to: obtain procedures not available in their home country, save money on procedures that are not covered by their government or private health insurance, and avoid long wait times for procedures in their home countries [4, 5, 6, 7]. Medical tourists from varied socio-economic backgrounds travel from a myriad of countries to obtain procedures in destination countries that are as diverse [1, 8, 9]. Another noteworthy feature of this globalizing health care industry is that friends and family members often accompany medical tourists abroad [1, 9, 10, 11, 12]. They may adopt a supportive role during the planning stages, while abroad, and upon return home, and are referred to here as caregiver-companions. In this article we shed light on caregiver-companions' central role as actors and stakeholders in medical tourism.

Little is known about caregiver-companions in medical tourism and the roles they take on in this global health services practice. In our previous research, it was found that they are an important part of many Canadian medical tourists' experiences at home and abroad [13]. Despite this, no dedicated research attention has been given to this group. The only published works that offer any depth of insight into the experiences of caregiver-companions are two biographical novels about medical tourists' journeys written by their caregiver-companions, though in both novels the authors focus heavily on the medical tourists' experiences and only somewhat on their own [14, 15, 16]. Some industry reports e.g. [17, 18] and academic studies (e.g. [11, 12, 13]) have also attempted to quantify the number of friends and family members who accompany medical tourists abroad, though these pieces do not offer an experiential perspective on caregiver-companions. Otherwise, we have little understanding of the roles played by caregiver-companions, the benefits and challenges they introduce as stakeholders in medical tourism, and the risks they may incur while abroad, among many other aspects of their experiences.

It is important to have a better understanding of caregiver-companions because they embody the informal caregiving (i.e., unpaid care by untrained individuals) component in the medical tourism industry. Research consistently shows that informal caregivers - usually patients' friends and family – play a key role in maintaining patients' health and wellness [19, 20, 21]. While a competent caregiver may improve a patient's health, one who is less prepared and supported may be detrimental to it [22, 23]. Providing informal care can also have a negative impact on the health of the caregiver. For example, informal caregivers are susceptible to caregiver burden due to stress and burnout, which can present as anxiety, depression, and other mental health issues [24, 25, 26, 27, 28,29, 30]. Having access to meaningful supports such as information or respite care providers, however, can assist with preventing the onset of such burden, and lead to a more positive caregiving experience [31, 32]. In this paper, we provide a glimpse into the unique practice of *transnational* informal caregiving in the context of medical tourism.

The existing literature has served to show that there are many actors or stakeholder groups that, together, enable the practice of medical tourism. Patients receive care while surgeons and other physicians provide it, facilitators aid with connecting patients to hospitals and making travel bookings, and policy-makers and government officials work to ensure that destination countries have adequate infrastructure to receive these bookings [33, 34]. International patient coordinators (IPCs) constitute yet another such group. They are professionals who work at or with medical tourism destination facilities to coordinate the on-site care of medical tourists. Among many other activities, they often create itineraries for patients and serve as a point of contact for questions or concerns. Their responsibilities typically begin before a patient's booking is secured and continue until after a patient has returned home. Here we examine informal caregiving in medical tourism from the perspective of IPCs. This allows us to obtain a breadth of insight, which we believe is important in order to identify issues that warrant subsequent investigation in depth. Out of any other medical tourism stakeholder group apart from the medical tourists themselves, IPCs have the most interaction with each caregiver-companion. In fact, it is usual for IPCs working in larger facilities to meet hundreds of caregiver-companions every year, which heightens their knowledge of trends among this medical tourism stakeholder group. It is for this reason that in the current article we examine facets of the caregiver-companion experience from the perspectives of IPCs. By interviewing IPCs from a range of medical tourism destination facilities across several countries, herein we are able to offer insights that have emerged from IPCs' interactions with many hundreds if not thousands of caregiver-companions and medical tourists alike.

In this article we present the findings of an analysis that examines the challenges that caregiver-companions can pose to IPCs, destination medical tourism hospitals/clinics and their staff, medical tourists, and even themselves from the perspectives of IPCs. It is important to note that the majority of IPCs we spoke with encouraged medical tourists to bring a friend or family member abroad with them because they commonly provide help and companionship that, they observed, positively affects medical tourist well-being. However, they also identified challenges that can be introduced with the presence of caregiver-companions. We believe that articulating these challenges helps to shed light on issues posed by the practice of informal caregiving in the global health services practice of medical tourism and the presence of caregiver-companions in destination facilities. To accomplish this, in the section that follows we outline the study design. Following this, we present the findings of a thematic analysis that identifies three characterizations of the challenges that caregiver-companions pose, in that they can: (1) require time, attention, and resources; (2) disrupt provisions of quality care; and (3) be exposed to risks. We then move to offer a discussion that contrasts the findings of the thematic analysis against existing findings in the medical tourism and informal caregiving literatures while also offering comments on directions for future research.

METHODS

This analysis is one component of a larger multi-method study. The purpose of the multi-method study is to gain an understanding of the experiences of Canadian medical tourists' caregiver-companions through gathering and analyzing their first-hand accounts and those of other stakeholders. Stakeholder groups consulted in the study are: Canadian medical tourists, Canadian medical tourists' caregiver-companions, Canadian medical tourism facilitators, and IPCs in various countries. In this paper, we focus exclusively on the IPC dataset. IPCs are an incredibly valuable informant group for the study because they interact routinely and closely with caregiver-companions, as well as medical tourists, and so are well placed to provide a breadth of insight into caregiver-companions' experiences. It is important to note that although the larger study to which this analysis contributes has a focus on outbound medical tourism by Canadians, the IPCs we interviewed were not asked to comment specifically on Canadian patients but, instead, spoke on broad trends about all the caregiver-companions with whom they have interacted.

Recruitment

Recruitment commenced upon receiving approval for the study from the Research Ethics Board at Simon Fraser University. Following receipt of this approval, we sought to recruit IPCs from a diverse range of countries and facilities to participate in phone and Skype interviews. Our target sample size was twenty, which is the number of IPCs we anticipated needing to speak with in order to achieve purposeful diversity in the sample. We used several methods to recruit participants, namely: (1) emailing letters of invitation to hospitals whose websites mentioned IPCs, IPCs found in online medical tourism directories, and IPCs who had posted on online forums; (2) snowballing out from existing participants; and (3) disseminating calls for participants through our team's networks and online forums and magazines. All emails and advertisements included the interviewer's contact information and indicated that interviews could be conducted in English or French (a request for a Spanish-language interview was also accommodated).

Upon receiving responses from potential participants, they were emailed a study information sheet and eligibility was confirmed. Eligibility required that participants were indeed IPCs and were based in or worked with medical tourism clinics or hospitals that offered surgical procedures that did not involve third parties (e.g., organ transplantation). Not all potential participants used the job title IPC, and so they had to indicate that in their jobs they were present in the facility with the medical tourist, made care and other arrangements, and assisted clients in a non-clinical capacity. In order to enable as much diversity among the sample to be captured, no more than 3 people from a single facility were interviewed. For those who met these eligibility requirements, an interview was scheduled at a time of their convenience.

Data collection

Telephone and Skype interviews were conducted from July to October, 2012. Interviews were semi-structured in order to ensure that core issues were consistently probed and to allow participants to introduce topics that captured unanticipated experiences and insights. To ensure consistency, the first author conducted all the interviews apart from one. That interview was with a Spanish-speaking participant and was conducted by a knowledgeable collaborator.

All phone interviews started by obtaining verbal consent, in which confidentiality to the furthest extent of the law was assured. Verbal consent was chosen over written consent to minimize the logistical difficulties inherent in the international scope of our study. Interviews generally lasted from 45 to 75 minutes. They covered a range of topics: (1) caregiver-companion characteristics, (2) interactions between caregiver-companions and hospital staff, (3) caregiver-companions' roles and responsibilities, and (4) the risks to which caregiver-companions can be exposed. Table 1 lists a sample of questions included in the interview guide. Data collection stopped when our target sample size was met, which coincided with the time our recruitment strategies ceased to generate interest from new potential participants.

Analysis

All interviews were recorded digitally and transcribed verbatim. Thematic analysis was employed, which involved six steps. First, the transcripts were review by all authors. Second, the team met face-to-face to discuss emerging themes as well as outliers. Third, themes that were collectively agreed upon were used to create a preliminary coding scheme by the first and second authors. The coding scheme identified umbrella concepts and their components into which data segments were categorized to inform thematic analysis [35]. Fourth, transcripts were coded by the first author using NVivo qualitative data management software, with input on code refinement being sought from the second author. Fifth, emerging trends and patterns relevant to the themes pursued in the current analysis were identified in the coded data by the first and second authors. Sixth, these trends and patterns were then compared to existing knowledge and the study objectives by the first and second authors and confirmed by the full team to reveal a refined interpretation of meaning in coded data [36]. Importantly, common themes emerged through this process despite differences in

Table 1. Selected interview questions.

Question	*Sub-probes*
What do you see as your role with regard to travel companions?	- How much interaction do you typically have with patients' travel companions?
	- What are some of the reasons that you interact with travel companions?
What kinds of responsibilities do you commonly see travel companions taking on?	- How prepared do you think travel companions are to take on these various responsibilities?
	- Are there any things you think could be done to assist with preparing them for these responsibilities?
What health and safety risks, including stressors, do travel companions face?	- Which of these risks has the most significant impact?
	- Who assists travel companions with minimizing these risks?

participant work environments and work histories, which is a characteristic of thematic analysis [37].

RESULTS

We conducted 20 interviews with 21 IPCs (one interview had two participants). The participants worked at 16 different facilities across Bolivia, Costa Rica, Barbados, Mexico, the United States, Croatia, India, Israel, Thailand and Turkey. Twelve of these facilities often received medical tourists from North American, six commonly saw Europeans, one saw mostly Australians, another saw mostly Africans, and two did not have a dominant region of origin for medical tourists. Some specialized in procedures such as cosmetic surgery, bariatric surgery, orthopedic surgery, oncology procedures, spinal surgeries, veinoplasty, and cardiac surgery, while others provided a myriad of procedures.

At the time of the interview, the length of time participants had been IPCs ranged from six months to 12 years. Five participants worked for companies that were based outside of the hospitals or clinics that they coordinated care for, 12 worked for independent facilities, and four worked for hospital chains. Six were the only IPC employed by their company or facility while 15 worked in IPC teams. The IPCs we interviewed took on a range of responsibilities towards medical tourists. For example, three commonly helped the patient determine an appropriate surgeon and facility, thirteen coordinated the itinerary (e.g., surgical date, days of hospital stay, local accommodations, etc.) and ground transportation (e.g., transport to and from the airport, local transport for the caregiver-companion), fourteen routinely provided education to the patient and the companion, two provided some limited nursing care to the patient, and all answered questions and addressed requests.

In the remainder of this section we discuss the three themes that emerged from the dataset regarding the challenges that IPCs, international facilities, and medical tourists can face when caregiver-companions accompany a medical tourist, namely that: (1) companions require time, attention and resources, (2) companions can disrupt the provision of quality care, and (3) companions can be exposed to risks. It should be noted that, according to the majority of IPCs, the help and companionship that caregiver-companions commonly provide outweigh these challenges and caregiver-companion accompaniment is therefore strongly recommended. We have included verbatim quotations throughout the section in order to enable the participants to 'speak' to these thematic findings. Following each quote we indicate the country in which the IPC participant was based as well as the number of years they had worked in this capacity.

Caregiver-companions can require time, attention, and resources

IPCs pointed out several ways in which the involvement of a caregiver-companion can serve to burden medical tourists. First, "*the costs are bigger when you travel with somebody of course, double airplane ticket, double accommodation, double meals, everything*" (Croatia, 2.5). In other words, there is a financial cost to bringing a caregiver-companion. Second, a caregiver-companion may unintentionally increase the time needed for recovery by creating additional stress for the patient. This can be a consequence of worry or anxiety because "*...if the [caregiver-companion] worries a lot then the patient is getting worried too and as you know the emotional state of the patient is really important for their recuperation process or their healing process...*" (Costa Rica, 0.6). Third, caregiver-companions can further be burdensome by doing nothing at all. In these cases, the companions accompany the patient to the destination country, but are otherwise absent from the medical tourism experience. One participant commented that some of "*the companions they feel that they are here on holiday so the patient is at the hospital themselves*" (Thailand, 6). Participants suspect that this neglectfulness is not intentional; rather, the companions do not understand the full scope of what is expected of them.

Although the patient is the focus of IPCs' responsibilities, in nearly all cases coordinators also reported the need to attend to caregiver-companions and in doing so divert some of their time and attention away from the medical tourist. For example, they provide guidance when needed. If the companion's presence is stressful to the patient, they may suggest a tourist activity: "*...in the event that we find that the patient is uncomfortable... we try to, to get... the companion to go somewhere [like]... an island tour or a nice restaurant*" (Barbados, 2). More often, IPCs will need to educate caregiver-companions regarding what to expect during the medical tourism process and what roles they should play. Education is an integral part of the participants' self-identified role towards the companions. It is often cited as a strategy to avoid stress from the medical tourist, the companion, and the facility staff: "*the more educated the companion, [the] better [it is] for all of us*" (Turkey, 1). The companion may initiate this education by asking questions. One participant noted that companions are *"always very, very worried, full of... maybe unnecessary questions... most often the companions of the patient are more worried and more, more, more curious than the patient himself"* (Croatia, 2.5). Some caregiver-companions also voice complaints about the facility, the doctor, or about the burden of their own worries. In the latter case, IPCs comfort them: *"...you're dealing with an emotionally charged individual to begin with and they have different expectations... and it takes a great deal of time and attention to calm them down"* (Thailand, 12). Like educating the companion, many IPCs saw offering comfort as strategically important to a positive experience and, although doing so took time, there were eventual benefits to be had from this time investment.

Caregiver-companions can create a redundancy of some roles with staff in medical tourism facilities. Many perform the hands-on care normally done by nurses, such as by helping the medical tourist with every day tasks after surgery while still in the facility. *"They assist the patient to get up and go to the bathroom, even though the nurses are available to do that... The same as showering, bathing, the nurse is completely capable of doing that but a lot of times the travel companion wants to do that instead"* (Mexico, 12). They also often help with recovery exercises, namely for mobility, and monitor the patient for symptom changes. The responsibility of offering comfort to the patient is sometimes adopted by the companion instead of facility staff because they are familiar and usually constantly at the patient's side. Caregiver-companions will also communicate with loved ones back home, a task that many IPCs are prepared to do. One participant said that *"some have their own GATT connection or Skype in their computers. And so [I say] 'I can make an international call for you'. And they say 'No, I don't need that. I brought my own equipment. I brought my Skype, I brought my jack, I brought this. I came prepared'"* (Costa Rica, 5). Although these redundancies are seemingly helpful in that they free up the time of facility staff, IPCs indicated that they can pose challenges to continuity in symptom monitoring as well as record keeping.

Caregiver-companions can disrupt the provision of quality care

Workers at medical tourism hospitals and clinics, and IPCs specifically, can find it challenging to accommodate companions because they do not always respect facilities' norms. For example, facility rules occasionally need to be changed in light of disruptions the companions introduce to the organizational routine. According to one participant, their facility needed to change the protocol regarding the number of visitors that patients could have in their room at once. The change occurred after the arrival of a patient's seven companions who "*overran the facility*" (USA, 8). Multiple participants noted that companions often complain about the food on behalf of the patient. In one case, a companion insisted that s/he prepare the patient's food in the facility's kitchen. As the participant recounts: "*we have even allowed the attendants of the companion to enter the kitchen and prepare their own kind of food with their own hands*" (India, 2.25). While it is the norm that caregiver-companions provide support and some hands-on care to the patient, in some instances this was not done. For example, in some cases the companions "*passively support*" (Thailand, 6) patients, meaning that they are physically present but they do not adopt any other caregiving responsibilities. Other companions are perceived as uncaring and absent from the bedside. These two types of disruption, exceptional caregiver-companion actions (e.g., insisting on preparing food for the patient) as well as unexpected inaction (e.g., offering only passive support), both create more work for facility staff and disrupt facility norms.

Facilities are sometimes faced with companions who enable patients to go against the advice given by staff and clinicians. Prior to arrival in the destination country, it is important for the patients to follow instructions in preparation for the surgery or to provide accurate information about their health status. This advice is, however, not always followed. Participants noted that in some cases caregiver-companions were complicit in this problematic behaviour. Then, in cases where physicians cancel the procedure as a result, companions can become upset (Thailand, 12):

I: Are there any exceptional or unusual cases of companions accompanying medical tourists that stick out in your mind?

P: Yes there are several, usually it's misinformation from the patient side. Patients who are HIV positive, who... [have] other underlying problems and they present for a simple procedure and you find out that they have a heart condition and many other things, there are major problems and you can't treat them... So you turn them away [after they have arrived].

I: ...How do the companions react to this?

P: ...Often outrage, because they've [caregiver-companions] been part [of it]...they [patients] think they can go to another country and get treatment for various things they've probably been turned down for in their own country.

Other caregiver-companions may believe that their competence is equal or superior to that of the nurses and physicians, and as a result do not act on clinical advice. This can present as over-protectiveness: "*sometimes the travel companion can be a little bit too over-protective and they think they know better than the doctor*" (Mexico, 1.5 & 0.75). They may also enable patients to ignore or act against facility rules such as those regarding smoking on the hospital or clinic grounds. Thus, caregiver-companions can enable patients to ignore rules and clinical instructions, and also avoid following them themselves.

Some IPCs reported on cases in which caregiver-companions were required to make decisions on clinical advice on behalf of the medical tourists. In the cases of caregiver-companions with limited financial resources, the financial affordability of medical care may inform decision-making and can override clinical advice and best practice, particularly in light of complications that require quick decisions and significant financial resources. One participant told the story of a caregiver companion he worked with (Turkey, 1.5):

Somebody's dad is in the ICU [intensive care unit] and...there were complications. You have to extend your dad's stay in the ICU, you don't know what's going to happen, and so cost just increases and increases. But you never really thought of ...such a long stay and such an expensive hospital particularly in the ICU which... without insurance is a tremendous cost. And so it was a very difficult situation for them I think both financially, emotionally: and 'how far do you go, how much is your dad worth?' that was really the question.

Caregiver-companions with limited financial resources may be faced with unanticipated financial challenges on top of the already debilitating cost of medical tourism.

Caregiver-companions can be exposed to risks

According to the IPCs we spoke with, caregiver-companions can be exposed to stress, which is a health risk, at a number of points throughout the course of a medical tour. The majority of participants noticed that before the patient has had surgery, companions typically experience stress from worry about the outcome of the procedure. "*I would say a little bit stressful... Once the patient's out and everything is okay they seem to be okay*" (Thailand, 1.5). Another reason for the stress is a lack of trust in the country, the facility, and/or the IPC experienced by some caregiver-companions. One participant summed it up as "*...the stress factors [are] the fears, the prejudice, judgments toward the country,* *towards the hospital, [and] treatment procedures*" (Turkey, 1). However, these stresses usually fade once caregiver-companions gain some understanding of the country and the hospital environment, and, according to many participants, a large part of the IPCs' interactions with the companions pre-surgery involves fostering this understanding. Participants also noted that during this time, IPCs often try to educate the companions on the procedure and care plan, as well as the ways in which the companions could be helpful to the patient. This helps to remove the stress associated with feeling unprepared, which is common amongst caregiver-companions who "*really have no clue what's happening*" (Turkey, 1.5). As one participant noted "*when the patient and the companion are informed about what's going on they're... they're more calm*" (Costa Rica, 2.5). Therefore, education and information are used to minimize or avoid stress.

After surgery, distinctive types of stresses are commonly experienced among caregiver-companions. Multiple participants reported emotional strain in companions due to worry for the patient, stress over an unknown outcome, and uncertainty of their role in assisting with the patients' pain management and overall wellbeing. One participant noted that "*sometimes [we] will actually have a companion that would be more on verge of a breakdown than the patient because they take such responsibility in the whole situation, wanting everything to be perfect, wanting the patient not to be in pain*" (Thailand, 6). It may also be difficult for caregiver-companions to handle the sight of swelling and surgical wounds immediately after a surgery, especially one that is cosmetic or bariatric in nature. However, the inability to handle this among caregiver-companions can create more stress for the patients and can slow their healing process. Therefore IPCs who work in facilities that offer cosmetic and bariatric procedures exclusively commonly advise patients to travel unaccompanied (Bolivia, 7):

Companions bring more stress to [patients]... Let's say, one of the stitches comes out early and you have never seen open stitches before, and... you can see inside the [patient] and they get really scared. So [companions] make the patient more stressed out than if [companions] are not there [with them].... We actually do not advise them to come with a companion.

Many participants characterized companions as worried, and therefore at risk of experiencing stress, because they are constantly anxious about the patient's welfare. Although the stresses may overlap, there are distinctions between pre- and post-surgery stresses. Before surgery, as discussed in the last paragraph, companions are concerned about trusting their environment as well as their own competencies, which can cause stress. After surgery, their focus tends to be strongly focused on the patients, which can lead to worry and thus mental stress and anxiety.

Companions can face safety risks, some of which can also threaten their health, while abroad. Participants noted that companions and medical tourists alike may be robbed, financially exploited, get lost in the city, or drink contaminated tap water. IPCs will sometimes educate caregiver-companions to about the potential travel risks to help mitigate them for the companion as well as the patient (Croatia, 2[illegible]5):

During the stay in Croatia they can be robbed... they can get lost in the different countries... the patient coordinator will instruct the companion what to do...how to behave in certain countries because the cultures are different, the things are different in each country... so I think the communication between the coordinator and the companion and then with the patient is crucial.

Risks associated with everyday life are also present while caregiver-companions are abroad. Participants recounted their experiences with companions who were in a bar fight, had fallen down a flight of stairs, were in a car accident, experienced a heart attack, and become ill with a stomach flu while abroad. One participant said: "*we had a companion who went on a hike and broke her leg*" (Israel, 3[illegible]). Given the range of risks that caregiver-companions may be exposed to while abroad, both within the facility and beyond, it is impossible for IPCs to anticipate the full scope of risks.

DISCUSSION

The findings shared above show that, according to IPCs who have spent many hours interacting with and addressing challenges posed by numerous caregiver-companions throughout their careers, caregiver-companions and the practice of informal caregiving can present multiple challenges to international medical tourism facilities, their staff, and even to medical tourists. Despite this, the IPCs we spoke with were overwhelmingly supportive of caregiver-companions accompanying medical tourists abroad, with the exception of some who worked with patients receiving bariatric or cosmetic surgeries, because caregiver-companions were perceived to be more helpful overall than challenging. However, they do present challenges. The IPCs pointed out that caregiver-companions may, for example, have a negative impact on the patient through cost of accompaniment or inadequate care provision. Caregiver-companions may also create unanticipated or extra work for IPCs and other workers in destination facilities, as additional clients and by ignoring established organizational rules, routines, and expectations. Furthermore, caregiver-companions may be susceptible to stresses and health and safety risks, which can further deteriorate their own abilities to offer the patient quality care. Our findings show that IPCs believe companions are more successful caregivers and less of a burden to staff when they are receptive to education from IPCs and willing to acting upon it. Our findings also show that the presence of a caregiver-companion can shift the IPC's focus from the patient to the companion at times. Building on these findings, in this section we contrast what this study has found about the challenges that caregiver-companions can pose against findings in the existing informal caregiving and medical tourism literatures. Doing so enables us to offer some insight into what is distinct about transnational informal caregiving in the medical tourism context.

Our analysis has revealed that caregiver-companions require additional resources. For example, medical tourists often shoulder the financial burden of an extra flight, accommodation and meals when they travel accompanied. For that reason, cost can be an issue of relevance to caregiver-companions' involvement in medical tourism, a point that was voiced by most IPCs we spoke with. Although other studies have shown that cost can be an important factor in medical tourists' decision-making [12, 38, 39, 40, 41, 42, 43] because it may be less costly to obtain a procedure abroad than at home [42, 44, 45], discussion of cost savings in medical tourism rarely mention the costs associated with bringing a caregiver-companion. In other words, the cost incentive associated with obtaining medical care abroad may be brought into question for those accompanied by a caregiver-companion; a point that is paralleled in some other medical tourism studies that have emphasized the heavy financial burden that can be placed on medical tourists and their families, and particularly those traveling from the Global South [9,46]. Our analysis further shows that caregiver-companions with less financial resources are equally motivated by cost when put in the position of having to make decisions based on clinical advice while abroad, especially when complications occur. This finding echoes concern in the existing medical tourism literature about the role that cost plays in medical tourists' making decisions based on clinical advice [47, 48, 49].

Our findings show that when caregiver-companions travel abroad, they may be exposed to risks and stresses derived from the unknown and unfamiliar nature of their surroundings. Caregiver-companions may even be apprehensive about the medical tourism facility and mistrust their standards of care. This is consistent with media reports of a popular schema about poor care quality and standards in developing countries [a][50, 51, 52, 53]. According to IPCs, such apprehension can lead caregiver-companions to disrupt a facility's organizational routines due to mistrust or misunderstanding, which can have significant implications for facility staff and even the medical tourist. Caregiver-companions may also hold pre-existing stereotypes or misconceptions about cultural norms that inform their apprehensions. This point is consistent with the findings of other studies that have pointed out that medical tourists may have concerns regarding the "foreignness" of the destination they are travelling to [16, 54,55]. Meanwhile, our study also shows that caregiver-companions' own apprehensiveness over "foreignness" and the unknown may lessen their abilities to address these same concerns held by medical tourist. Related to this, the unfamiliar care environment is a key difference between the contexts of caregiving in medical tourism and more conventional informal caregiving at home. Conventional caregivers are not burdened by the apprehension of "foreignness", nor are they predisposed to doubt the quality of local facilities where they spend a minority of their time as caregivers. The need to travel to a foreign facility introduces additional challenges for caregiver-companions that can bring about new stresses or amplify existing ones.

The findings of our study show that caregiver-companions are often needy, in that they require time, attention, and resources, and that IPCs attempt to address these needs. IPCs are available by telephone while the companions and patients are abroad, and they are often physically present while the patient is in the hospital to assist with meeting these needs. While the support, advice, and information they provide to caregiver-companions may seemingly present as extra work, the IPCs that we spoke with viewed it as an intervention aimed at minimizing stress, burden, and other negative outcomes for caregiver-companions, the medical tourist, and facility staff. This aspect of the IPC's role, in relation to supporting the caregiver-companion, is consistent with the findings of other studies that have shown that medical tourism facilities tend to have significant staffing resources and provide a high amount of interpersonal time dedicated to patients [11, 56]. Attentiveness towards caregiver-companions by IPCs can take the form of providing guidance, education, and comfort in order to be responsive to their needs. The provision of education in particular may assist with overcoming what is often reported to be the lack of reliable information available to those engaging in medical tourism [57, 58]. Although information being shared by an IPC as an educational strategy to meet caregiver-companions' needs and avoid the onset of stress is unlikely to be neutral, a concern raised about the commonness of industry-generated information being used to inform decisions about medical tourism [46, 58, 59], the reportedly trusting relationship built between the companion and coordinator may enhance its perceived trustworthiness.

Multiple studies have shown that having access to meaningful formal support for informal caregivers, such as that provided by IPCs in the context of medical tourism, can translate into increased health and wellbeing for the patient [60, 61] and caregiver alike [24, 25, 26, 62]. The findings of our research confirm that, in the experience of our participants, support in the form of dedicated collaboration between the companion and the IPC is necessary to improve patient outcomes and avoid a slowed recovery. However, researchers and policy-makers have been quick to point out that there is a difference between having access to support and the uptake of this same support, and that some informal caregivers are simply not accepting of assistance in any form [63, 64]. IPCs seem to be cognizant of this and they offer support actively and intentionally as a strategy to elucidate a better experience for the medical tourist, the companion, and hospital staff. This suggests that caregiver-companions in medical tourism may be equally in need of being prompted to accept support in order to minimize their stress and burden as other types of informal caregivers.

Informal caregivers typically experience more stress when they are not able to effectively collaborate with health professionals [60, 64, 65, 66, 67]. Similarly, our findings show that when caregiver-companions are thought to be not open to collaborating with medical tourism facility staff, and most specifically the IPC, it is more difficult for them to carry out their caregiving responsibilities and they may find the experience more stressful. Participants, for example, reported that companions can be incapable of or ineffective at helping patients if they are too worried about surgical outcomes or their own care abilities. Worry is also a sign of caregiver burden [26, 68, 69] and, not surprisingly, some IPCs identified it as a potential source of stress and negative health outcomes for caregiver-companions. Our findings have also shown that worry is very common throughout the companion's time abroad and can sometimes negatively impact the patient who, according to other medical tourism studies, may already be experiencing stress [55, 70, 71].

Directions for future research

While our analysis sheds new and dedicated light on the challenges introduced by caregiver-companions in medical tourism, and also on the roles and responsibilities more generally associated with this stakeholder group, we believe that much research remains to be done about the intersection between medical tourism and informal caregiving. Our findings have shown some differentiation between caregiver-companions as a result of socio-economic status and the procedure the patient is obtaining abroad, but it is unclear if there are meaningful differences between them based on their relationship with the medical tourist, their country of origin, or the destination country, and what these differences may be. Meanwhile, some of these differences have been found to be extremely important to caregiver health status and patient outcomes in other forms of informal caregiving [72, 73]. Our analysis also suggests that the presence of caregiver-companions and the actions they undertake can be beneficial to the health of the patient, and that support from IPCs is beneficial to caregiver-companions and thus ultimately the patient. However, we require more knowledge regarding the other forms of support caregiver-companions draw from or that they require. For example, would caregiver-companions benefit from gaining access to some or all of the interventions developed for informal caregivers in various health systems and governments, such as system navigation tools, support groups, and therapeutic interventions [31, 74, 75]? This is a pressing question that researchers can assist with answering.

This analysis helps to demonstrate the value in speaking with stakeholders in medical tourism who are knowledgeable about other stakeholder groups, and particularly groups that are difficult to locate (as is the case with caregiver-companions). In particular, it highlights the potential for having IPCs serve as participants in medical tourism research. Our findings have shown IPCs' perspectives on the challenges introduced by caregiver-companions, but a counter-point to this, examining the benefits they introduce for IPCs and other destination facility staff would be a useful direction for future research. Our findings have further shown that the extra expense of travelling abroad with a friend or a family member may discourage some medical tourists from taking a companion. However, research has hinted at the regularity with which friends and family members accompany medical tourists abroad

[9, 10, 11]. A detailed account of the reasons for which they are perceived as useful enough to justify the extra expense has yet to be undertaken. IPCs would be ideal participants for such a study for the same reasons that we found them helpful for our own: they have the most interaction with the highest number of companions out of any stakeholder group that we are aware of and they can therefore speak to the breadth and depth of a particular issue relevant to those they interact with in a professional capacity.

Limitations

First, recruitment information indicated that interviews could be conducted in French and English. Upon request of an interested potential participant, we were able to coordinate a Spanish-language interview as well. IPCs who were not fluent in these languages were not included in this study. Second, the utilization of semi-structured phone and Skype interviews may have resulted in some missed data due to the inaccessibility of visual cues and the formality of the conversation. Studies have shown, however, that phone interviews are a reliable and cost effective method for data collection [76], and for this reason we are confident in the soundness of the dataset. Third, in this analysis we have considered many aspects of the caregiver-companion's experience in medical tourism, yet we did not consult with a single caregiver-companion in doing so. While we believe that our use of IPCs to gather information about this stakeholder group is well justified, it is nonetheless a limitation that we present information about caregiver-companions here solely from the perspective of IPCs. A clear direction for future research, including our own, is to speak with caregiver-companions themselves about the experiences embedded in the current analysis along with other aspects of their informal caregiving roles and responsibilities.

CONCLUSIONS

In this article, we presented the findings of a thematic analysis derived from 20 interviews with 21 IPCs working in 16 medical tourism facilities across nine countries. We examined concerns that these IPCs identified regarding the challenges that caregiver-companions and the practice of informal caregiving may pose to facility staff and medical tourists. These concerns include the potential for caregiver-companions to be exposed to health and safety risks that negatively impact not only themselves but also the medical tourist. It was also shown that caregiver-companions can be burdensome to medical tourism facility staff, including the IPCs, through requiring time and attention in order to minimize or lessen anxiety, worry, and ultimately stress. IPCs tend to actively take steps to mitigate the potential for challenges to arise in relation to informal caregiving in medical tourism, with a particular focus on caregiver-companion education. Our findings suggest that if companions are open to collaboration with IPCs, and particularly in the form of information sharing, then their experience abroad can be safer and less stressful for themselves and, by extension, for the accompanied patients and facility staff.

This article has positioned caregiver-companions as a central stakeholder group in the global health services practice of medical tourism and has also shown the value of asking one stakeholder group, namely IPCs, about their experiences with another, namely caregiver-companions, to gain broader knowledge. Most broadly, the analysis shed light on the very practice of informal caregiving in medical tourism, which is a particular form of transnational caregiving – a form of caregiving that has received little dedicated research attention. Caregiver-companions represent the intersection between medical tourism and informal caregiving and, to our knowledge, this is the first account that focuses on this intersection. We encourage more research to be done so that informal caregiving in medical tourism can be further examined and also so that appropriate responses to and interventions for this transnational caregiving practice can be implemented.

ENDNOTE

[a]Medical tourists from developing countries might perceive the opposite: the local standard of care might be seen as inferior whereas the foreign one is held on a pedestal [9].

ACKNOWLEDGEMENTS

We would like to thank those who participated in this study for sharing their experiences and views with us. Thank you to Alejandro Ceron for conducting the Spanish language interview. The study was funded by an Operating Grant from the Canadian Institutes of Health Research (funding reference #115692). VAC is also funded by a Scholar Award from the Michael Smith Foundation for Health Research.

COMPETING INTERESTS

The authors declare that they have no competing interests.

AUTHORS' CONTRIBUTIONS

VC conducted the interviews, led the coding, and played a leadership role in writing this article. VAC is lead investigator on the study that funded this research and developed the protocol. She contributed to all aspects of the study and was heavily involved in preparing this manuscript. JS and LT contributed input to the data collection instrument, participated actively in the face-to-face analysis meeting, and provided feedback on drafts of this paper. JS also aided in recruitment. All authors reviewed and approved the submitted manuscript.

REFERENCES

1. Connell J: Contemporary medical tourism: Conceptualisation, culture and commodification. Tourism Manage. 2013, 34: 1-13.
2. Lunt N, Smith R, Exworthy M, Green ST, Horsfall D, Mannion R: Medical tourism: Treatments, markets and health system implications: A scoping review. OECD. 2011, http://www.oecd.org/els/health-systems/48723982.pdf,
3. Ehrbeck T, Guevara C, Mango PD, Cordina R, Singhal S: Health care and the consumer. McKinsey Quart. 2008, 4: 80-91.
4. Mattoo A, Rathindran R: How health insurance inhibits trade in health care. Health Aff (Millwood). 2006, 25: 358-368.
5. Carrera P, Bridges J: Globalization and healthcare: understanding health and medical tourism. Expert Rev Pharmacoecon Outcomes Res. 2006, 6: 447-454.

6. Eggertson L: Wait-list weary Canadians seek treatment abroad. Can Med Assoc J. 2006, 174: 1247-

7. Milstein A, Smith M: America's new refugees – seeking affordable surgery offshore. New Engl J Med. 2006, 355: 1637-40.

8. Hopkins L, Labonte R, Runnels V, Packer C: Medical tourism today: What is the state of existing knowledge. J Public Health Pol. 2010, 31: 185-198.

9. Kangas B: Hope from abroad in the international medical travel of Yemeni patients. Anthropol Med. 2007, 14: 293-305.

10. Yeoh E, Othman K, Ahmad H: Tourism Management Understanding medical tourists: Word-of-mouth and viral marketing as potent marketing tools. Tourism Manage. 2013, 34: 196-201.

11. NaRanong A, NaRanong V: The effects of medical tourism: Thailand's experience. Bull World Health Organ. 2011, 89: 336-344.

12. Yu JY, Ko TG: A cross-cultural study of perceptions of medical tourism among Chinese, Japanese and Korean tourists in Korea. Tourism Manage. 2012, 33: 80-88.

13. Crooks VA, Snyder J, Johnston R, Kingsbury P: Perspectives on Canadians' involvement in medical tourism: Final research report. [http://www.sfu.ca/medicaltourism/Perspectives_on_Canadians_Involvement_in_Medical_Tourism.pdf]

14. Kingsbury P, Crooks VA, Snyder J, Johnston R, Adams K: Narratives of emotion and anxiety in medical tourism: on *State of the Heart* and *Larry's Kidney*. Soc Cult Geogr. 2012, 13: 361-378.

15. Rose DA: Larry's Kidney: Being the True Story of How I Found Myself in China with My Black Sheep Cousin and His Mail-Order Bride, Skirting the Law to Get Him a Transplant–and Save His Life. 2009, New York: William Morrow

16. Grace M: State of the Heart: A Medical Tourist's True Story of Lifesaving Surgery in India. 2007, Oakland, CA: New Harbinger Publications

17. Medical Tourism Association: Patient Survey. [http://www.medicaltourismassociation.com/userfiles/files/Patient%20Survey-FINAL.PDF]

18. Travers R, Thelen S, Helmy E, Nabil N, Elmasri AW: Medical tourism development strategy final report. [http://www.imcegypt.org/studies/FullReport/Medical%20Tourism%20Development%20Strategy_EN.pdf]

19. Molloy GJ, Johnston DW, Witham MD: Family caregiving and congestive heart failure: Review and analysis. Eur J Heart Fail. 2005, 7: 592-603.

20. Bevan JL, Pecchioni LL: Understanding the impact of family caregiver cancer literacy on patient health outcomes. Patient Educ Couns. 2008, 71: 356-364.

21. Wright AA, Keating NL, Balboni TA, Matulonis UA, Block SD, Prigerson HG: Place of death: Correlations with quality of life of patients with cancer and predictors of bereaved caregivers' mental health. J Clin Oncol. 2010, 28: 4457-4464.

22. Cook JA, Cohler BJ, Pickett SA, Beeler JA: Life-Course and Severe Mental Illness: Implications for Caregiving Within the Family of Later Life. Fam Relat. 1997, 46: 427-436.

23. Mittelman MS, Haley WE, Clay OJ, Roth DL: Improving caregiver well-being delays nursing home placement of patients with Alzheimer disease. Neurology. 2006, 67: 1592-1599.

24. Carretero S, Garces J, Rodenas F, Sanjose V: The informal caregiver's burden of dependent people: Theory and empirical review. Arch Gerontol Geriat. 2009, 49: 74-79.

25. Hawranik PG, Strain LA: Giving voice to informal caregivers of older adults. Can J Nurs Res. 2007, 39: 157-172.

26. Rocha Pereira H, Rebelo Botelho MA: Sudden informal caregivers: the lived experience of informal caregivers after an unexpected event. J Clin Nurs. 2011, 20: 2448-2457.

27. Lee H, Singh J: Appraisals, Burnout and Outcomes in Informal Caregiving. Asian Nurs Res. 2010, 4: 32-44.

28. Borg C, Hallberg IR: Life satisfaction among informal caregivers in comparison with noncaregivers. Scan J Caring Sci. 2006, 20: 427-438.

29. Pinquart M, Sorensen S: Differences between caregivers and noncaregivers in psychological health and physical health: A meta-analysis. Psychol Aging. 2003, 18: 250-267.

30. Roth DL, Perkins M, Wadley VG, Temple EM, Haley WE: Family caregiving and emotional strain: Associations with quality of life in a large national sample of middle-aged and older adults. Qual life res. 2009, 18: 679-688.

31. Commission on the Future Health Care in Canada: Building on values: The future of health in Canada. [http://publications.gc.ca/collections/Collection/CP32-85-2002E.pdf]

32. Friedman B, Wamsley BR, Liebel DV, Saad ZB, Eggert GM: Patient satisfaction, empowerment, and health and disability status effects of a disease management: Health promotion nurse intervention among medicare beneficiaries with disabilities. Gerontologist. 2009, 49: 778-792.

33. Martínez Álvarez M, Chanda R, Smith RD: The potential for bilateral agreements in medical tourism: A qualitative study of stakeholder perspectives from the UK and India. Global Health. 2011, 7: 11-

34. Birch DW, Vu L, Karmali S, Johnson Stoklossa C, Sharma AM: Medical tourism in bariatric surgery. Am J Surg. 2010, 199: 604-608.

35. Schwandt TA: Thematic analysis. The SAGE Dictionary of Qualitative Inquiry. 2007, Thousands Oaks, CA: SAGE Publications Inc, 292-3

36. Fereday J, Muir-Cochrane E: Demonstrating rigor using thematic analysis: A hybrid approach of inductive and deductive coding and theme development. Int J Qual Meth. 2006, 5: 80-92.

37. Braun V, Clarke V: Using thematic analysis in psychology. Qual Res Psychol. 2006, 3: 77-101.

38. Leng CH: Medical tourism in Malaysia: international movement of healthcare consumers and the commodification of healthcare. Asia Res Inst. 2007, 83: 1-32.

39. Horowitz MD, Rosensweig JA: Medical tourism: health care in the global economy. Phys Exec. 2007, 33: 24-30.

40. Garud AD: Speaking for ourselves: Medical tourism and its impact on our healthcare. Natl Med J India. 2005, 18: 318-319.

41. Lindsay G: Medical leave. Fast Comp. 2008, 125: 108-119.

42. Glinos IA, Baeten R, Boffin N: Cross-border contracted care in Belgium hospitals. Patient mobility in the European Union: learning from experience. Edited by: Rossenmöller M, McKee M, Baeten R. 2006, Copenhagen, Denmark: European Observatory on Health Systems and Policies, 97-118.

43. Cohen IG: Protecting patients with passports: Medical tourism and the patient protective- argument. Iowa Law Rev. 2010, 95: 1467-1567.

44. The Gallup Organization: Cross-border health services in the EU. [http://ec.europa.eu/public_opinion/flash/fl_210_en.pdf]

45. Burkett L: Medical tourism: Concerns, benefits, and the American legal perspective. J Legal Med. 2007, 28: 223-245.

46. Johnston R, Crooks VA, Snyder J: "I didn't even know what I was looking for": A qualitative study of the decision-making processes of Canadian medical tourists. Global Health. 2012, 8: 23-

47. Cheung IA, Wilson A: Arthroplasty tourism. Med J Aust. 2007, 187: 666-667.

48. Turner L: 'First world health care at third world prices': Globalization, bioethics and medical tourism. Biosocieties. 2007, 2: 303-325. 1

49. Turner L: (2007b) Medical tourism: Family medicine and international health-related travel. Can Fam Physician. 2007, 53: 1639-1641.

50. Casacchia C: Medical tourism' gains steam as health care costs rise in US. [http://www.bizjournals.com/phoenix/stories/2010/05/03/story7.html]

51. Meyer B: Medical tourism: Lured by stem cells' promise, Americans head abroad for medical treatment. [http://www.cleveland.com/world/index.ssf/2009/02/medical_tourism_lured_by_promi.html]

52. Russell R: Surgery in the sun lures patients to Thailand. [http://uk.reuters.com/article/2007/03/07/uk-thailand-medical-idUKSP13935820070307]

53. Martin R: Clark Howard Discusses His Cancer Diagnosis With Dr. Randy Martin. [http://www.wsbtv.com/news/news/clark-howard-discusses-his-cancer-diagnosis-with-d/nFBhr/]

54. Mudur G: Hospitals in India try and woo patients. BMJ. 2004, 328: 1338-1338.

55. Howze KS: Medical tourism: symptom or cure?. Georgia L Rev. 2007, 41: 1013-1052.

56. Connell J: Medical tourism: Sea, sun, sand and ... surgery. Tourism Manage. 2006, 27: 1093-1100.

57. Lunt N, Carrera P: Medical tourism: Assessing the evidence on treatment abroad. Maturitas. 2010, 66: 27-32.

58. Lunt N, Hardey M, Mannion R: Nip, tuck and click: Medical tourism and the emergence of web-based health information. Open Med Inform J. 2010, 4: 1-11.

59. Eysenbach G: An ontology of quality initiatives and a model for decentralized, collaborative quality management on the (semantic) world wide web. J Med Internet Res. 2001, 3: 34-10.2196/jmir.3.4.e34.

60. Northouse L, Williams AI, Given B, McCorkle R: Psychosocial care for family caregivers of patients with cancer. J Clin Oncol. 2012, 30: 1227-1234.

61. Surbone A, Baider L: Are oncologists accountable only to patients or also to their families? An international perspective. 2012, Alexandria, VA: American Society of Clinical Oncology

62. Mangan PA, Taylor KL, Yabroff KR, Fleming DA, Ingham JM: Caregiving near the end of life: Unmet needs and potential solutions. Palliat Support Care. 2003, 1: 247-259.

63. Bass D, Noelker LS, Rechlin LR: The moderating influence of service use on negative caregiving consequences. Gerontol B Psychol Sci Soc Sci. 1996, 51B: S121-S131.

64. Weinberg DB, Lusenhop RW, Hoffer Gittell J, Kautz CM: Coordination between formal providers and informal caregivers. Health Care Manage Rev. 2007, 32: 140-149.

65. Bucher JA, Loscalzo M, Zabora J, Houts PS, Hooker C, BrintzenhofeSzoc K: Problem-solving cancer care education for patients and caregivers. Cancer Pract. 2001, 9: 66-70.

66. Morse JM, Pooler C: Patient-family-nurse interactions in the trauma-resuscitation room. Am J Crit Care. 2002, 11: 240-249.

67. Egerod I, Overgaard D: Taking a back seat: support and self-preservation in close relatives of patients with left ventricular assist device. Eur J Cardiovasc Nurs. 2012, 11: 380-387.

68. Jacobs B: From sadness to pride: seven common emotional experiences of caregiving. Always on Call: When Illness Turns Families into Caregivers. Edited by: Levine C. 2004, Nashville: Vanderbilt University Press, 111-125.

69. Wolpe G: (2004) A crisis of caregiving, a crisis of faith. Always on Call: When Illness Turns Families into Caregivers. Edited by: Levine C. 2004, Nashville: Vanderbilt University Press, 34-44.

70. Law J: Sun, sand and stitches. Profit. 2008, 27: 69-70.

71. Lautier M: Export of health services from developing countries: The case of Tunisia. Soc Sci Med. 2008, 67: 101-110.

72. Janevic MR, Connell CM: Racial, Ethnic, and Cultural Differences in the Dementia Caregiving Experience. Gerontologist. 2001, 41: 334-347.

73. Dentinger E, Clarkberg M: Informal caregiving and retirement timing among men and women: Gender and caregiving relationships in late midlife. J Fam Issues. 2002, 23: 857-879.

74. Rizzo VM, Gomes A, Chalfy A: Using an aging services agency/university partnership to study informal caregiver respite programs. Ageing Int. 2013, 38: 71-81.

75. World Federation for Mental Health: Caring for the caregiver: Why your mental health matters when you are caring for others. [http://www.wfmh.org/PDF/Caring%20for%20the%20Caregiver%2011_04_09%20FINAL%20(3).pdf]

76. Laws S, Harper C, Marcus R: Telephone interviews. Research for Development. 2003, Thousand Oaks, CA: SAGE, 298-299.

8

CHAPTER

Knowledge Brokers, Companions, and Navigators: A Qualitative Examination of Informal Caregivers' Roles in Medical Tourism

Victoria Casey[1], Valorie A Crooks[1*], Jeremy Snyder[2] and Leigh Turner[3]

[1]*Department of Geography, Simon Fraser University, Burnaby, British Columbia, Canada.*

[2]*Faculty of Health Sciences, Simon Fraser University, Burnaby, British Columbia, Canada.*

[3]*Center for Bioethics, School of Public Health, and College of Pharmacy, University of Minnesota, Minneapolis, Minnesota, USA.*

ABSTRACT

Introduction

Many studies examining the phenomena of medical tourism have identified health equity issues associated with this global health services practice. However, there is a notable lack of attention in this existing research to the informal care provided by the friends and family members who typically accompany medical tourists abroad. To date, researchers have not examined the care roles filled by informal caregivers travelling with medical tourists. In this article, we fill this gap by examining these informal caregivers and the roles they take on towards supporting medical tourists' health and wellbeing.

Methods

We conducted 21 interviews with International Patient Coordinators (IPCs) working at medical tourism hospitals across ten countries. IPCs work closely with informal caregivers as providers of non-medical personal assistance, and can therefore offer broad insight on caregiver roles. The interviews were coded and analyzed thematically.

Results

Three roles emerged: knowledge broker, companion, and navigator. As knowledge brokers, caregivers facilitate the transfer of information between the medical tourist and formal health care providers as well as other staff members at medical tourism facilities. The companion role involves providing medical tourists with physical and emotional care. Meanwhile, responsibilities associated with handling documents and coordinating often complex journeys are part of the navigation role.

Conclusions

This is the first study to examine informal caregiving roles in medical tourism. Many of the roles identified are similar to those of conventional informal caregivers while others are specific to the transnational context. We conclude that these roles make informal caregivers an integral part of the larger phenomenon of medical tourism. We further contend that examining the roles taken on by a heretofore-unconsidered medical tourism stakeholder group sheds valuable insight into how this industry operates and that such knowledge is necessary in order to respond to the health equity debates that surround this particular global health services practice.

Keywords: Informal caregiving Medical tourism Global health services Healthcare providers Qualitative analysis Family caregivers

INTRODUCTION

Health care delivery involves many provider groups, both formal and informal, that address different facets of patient care. Doctors and nurses are examples of formal provider groups in that they receive specialized training specific to the care they deliver, are paid to deliver this care, and are commonly licensed professionals [1, 2]. Much social health research has documented the important and complex responsibilities that formal providers assume in ensuring patients' health and general wellbeing e.g. [3, 4, 5]. In contrast, informal health care providers include friends, family, and some volunteers. Members of these groups provide essential care despite the fact that they are not formally trained and paid health professionals [6, 7]. These informal care providers, or caregivers, often provide ongoing essential care in the home, such as administering medications, managing wounds, and assisting with rehabilitation [8]. Meanwhile in hospitals and other types of residential care settings, they perform tasks that augment services provided by on-site formal providers. Examples of such tasks include monitoring symptoms and articulating patients' preferences to health care professionals [9, 10, 11]. The roles that formal and informal providers fill and the responsibilities assigned to each collectively work together to enable health care delivery. In this article, we examine the caregiving roles assumed by a specific group of informal health care providers: the friends and family members that accompany medical tourists abroad for private medical care.

Hospitals, clinics, and individual health care practitioners who provide medical treatment to privately-paying traveling international patients who do not have a formal cross-border care referral are part of what is often described as the 'medical tourism industry' [12, 13, 14]. This industry is reported to be on a multi-billion dollar scale and involves the transnational movement of patients to a growing list of destination countries located on nearly every continent [13, 15, 16, 17, 18]. There are a great variety of procedures that can be obtained in medical tourism facilities, including, though not limited to, cosmetic surgeries, cardiac surgeries, and orthopedic surgeries [15, 19,20]. Significant concern has been raised that the medical tourism industry is exacerbating health inequities in destination countries through such measures as recruiting formal health care providers out of public systems and into private ones, which prices locals out of needed health care [21]. Supporters of the industry meanwhile suggest that the practice of treating privately-paying international patients can assist with addressing health inequities through measures such as retaining formal providers who may otherwise have left the country in search of higher pay or more technologically sophisticated practice and bringing in capital [21].

Research examining medical tourism has paid some consideration to numerous health care provider groups central to this global health service practice. Perhaps not surprisingly, much of this research examines formal health care providers rather than informal ones, though sometimes members of the latter category are mentioned in passing. Although academic studies and industry reports note that it is common for medical tourists to be accompanied by friends and family members (e.g., [6, 16, 22, 23, 24]), they are rarely framed as informal health care providers or as a stakeholder group meriting additional exploration. Meanwhile, it is widely acknowledged that informal caregivers provide the bulk of patient care throughout the life course and across the care continuum and play an important role in establishing care recipients' health outcomes [7, 24, 25, 26]. Given the importance of informal caregivers in health care provision, it is surprising that their contributions to the care of medical tourists have not attracted greater scrutiny. Only non-academic work has addressed the practice of informal caregiving in medical tourism. Two published narratives of medical tourists' journeys written by their informal caregivers reveal that significant caregiving responsibilities were assumed by these individuals [27, 28]. Their detailed narratives document numerous instances of making decisions on behalf of the patient, liaising with formal providers, coordinating appointment scheduling, offering hands-on care, providing emotional and spiritual support, and taking responsibility for managing care-related finances. These narratives, although not scholarly in nature, are nonetheless illuminating because they suggest that informal providers can play a significant role in the practice of medical tourism. In this article, we aim to identify and explicate the informal caregiving roles that friends and family members assume in the course of a medical tourist's journey. By 'roles' we refer to their key social functions in the maintenance of medical tourists' health and wellbeing and the larger transnational care practice of medical tourism.

Our exploration of informal caregivers' roles herein is informed by insights gleaned from interviews with International Patient Coordinators (IPCs) working at medical tourism facilities. IPCs work at destination facilities; their task is to coordinate medical tourists' care. Their responsibilities include arranging ground transportation and local travel, communicating with doctors, scheduling medical appointments, and providing support and guidance for patients and their caregivers. Because of the nature of their jobs, every year they interact with anywhere from tens to hundreds of medical tourists and their informal caregivers. Given their function, we believe that by sharing their observations and experiences they are well positioned to identify the informal care roles filled by this caregiver group. In the section that follows we provide an overview of the study design and a description of the 21 IPCs with whom we spoke. We then present the findings of a thematic analysis that identified three roles commonly filled by medical tourists' informal caregivers: knowledge broker, companion, and navigator. We subsequently discuss the findings in light of the existing medical tourism and informal caregiving literatures and offer directions for future research. We conclude by reflecting on the relevance of this analysis for providing new insights that have relevance for the health equity debates that surround the global medical tourism industry.

METHODS

This analysis emerges from a large, multi-method study that explores first-hand accounts of medical tourists' informal caregivers and those who have worked closely with them in a professional capacity. Here, we report on the findings of interviews conducted with IPCs about their interactions with and observations of these caregivers. The findings speak to the roles that caregivers from a range of home countries fill as they accompany medical tourists seeking a variety of medical procedures at international health care facilities.

IPC recruitment commenced upon receiving approval from the Research Ethics Board at Simon Fraser University. We sought participants from a diverse range of countries and facilities using several concurrent methods: (1) emailing letters of invitation to hospitals and clinics whose websites mentioned IPCs, IPCs identified in online medical tourism directories, and IPCs who had posted on online forums; (2) snowballing out from initial participants; and (3) disseminating calls for participants through our team's networks and online medical tourism industry forums and magazines. Recruitment materials indicated that interviews could be conducted in English or French. A later request for a Spanish-language interview was also accommodated.

Interested potential participants who contacted us by e-mail were sent an information sheet that provided additional information concerning the study and described their rights as participants including confidentiality. Before this sheet was sent, participant eligibility was confirmed. Because many potential participants did not use 'IPC' as their official job title, they were required to indicate that: they worked with international patients who obtained procedures at medical tourism hospitals or clinics that offered surgical procedures without third party involvement such as organ transplantation; they were physically present in the facility with the medical tourist; they made care and other

Table 1. Selected interview questions

Question	Sub-probes
In your experience, what is the typical relationship between patients and their travel companions?	What are some of the common characteristics of travel companions?
What are some of the reasons that you interact with travel companions?	Before arrival? While abroad? Upon returning home?
What kinds of responsibilities do you commonly see travel companions taking on?	In relation to: communication; symptom monitoring; hands-on care; corresponding with friends and family at home; providing emotional support to the patient; providing spiritual support to the patient; making arrangements; travel and tourism activities; other relevant activities?
Have you ever experienced a situation where a travel companion's health worsened or improved while they were abroad (both in hospital or after discharge)?	How common is this? What could be the cause of their worsened or improved health? Can you think of any problems, stresses, difficulties that travel companions face while the patient is in hospital and after discharge?

arrangements; and they assisted clients in a non-clinical capacity. To capture diversity among the sample, no more than three individuals from a single facility were interviewed. We stopped active recruitment and interviewing when we reached our target sample size of 20 interviews, a point that coincided with when new potential participants were no longer being identified.

Interviews were conducted over telephone or Skype according to the participant's preference. They typically lasted for 45-75 minutes. A semi-structured interview guide was used, enabling the capture of issues central to the study's objectives and topics that were important to the participants. The first author conducted 19 English-language interviews while a knowledgeable collaborator conducted one in Spanish. No French-language interviews were requested.

Verbal consent was obtained before each interview, and the interviews lasted for approximately 45-60 minutes. As shown in Table 1, the interviews covered topics such as: (1) informal caregiver characteristics, (2) interactions between caregivers and medical tourism facility staff, (3) caregivers' roles and responsibilities, and (4) the risks to which caregivers can be exposed while travelling with medical tourists and providing care to them.

Twenty interviews were conducted with 21 IPCs (one interview had two participants) and drew from their experiences working at 16 different medical tourism hospitals or clinics in Bolivia, Costa Rica, Barbados, Mexico, the United States, Croatia, India, Israel, Thailand, and Turkey. Twelve IPCs mainly dealt with North American medical tourists, six mostly serviced Europeans, one primarily saw Australians and another Africans, and the remaining two did not report a particular regional orientation. The procedures provided at the facilities where the participants worked included cosmetic surgery, bariatric surgery, orthopedic surgery, oncology procedures, spinal surgeries, veinoplasty, and cardiac surgery.

All interviews except one were recorded digitally and transcribed verbatim. Technical difficulties prevented the exception from being recorded, and detailed interviewer notes were instead used to document the interview. All transcripts and notes were loaded into NVivo, a qualitative data management program, after which thematic analysis was conducted.

The thematic analysis involved six steps. First, all investigators reviewed the transcripts and notes. Second, emerging themes and outliers were identified during a face-to-face meeting with all investigators. Third, the first and second authors created a preliminary coding scheme that identified overall thematic concepts and their components. This involved collaboratively creating tiers of category headings that were manipulated until the authors were confident that they would entirely capture the study objectives as well as the themes and outliers identified in the meeting. Fourth, the first author coded the data in NVivo, with input on code refinement and interpretation from the second author. Fifth, the first and second authors identified emerging trends and patterns relevant to the themes pursued in the current analysis, namely those pertaining to caregivers' roles. Sixth, a refined interpretation of meaning in the coded data was revealed through a comparison of the trends and patterns with existing knowledge and the study objectives [29]. This comparison was initially done by the first and second authors and then confirmed by the full team. Characteristic of thematic analysis, this analytical process enabled common themes to emerge despite the differences in IPC participants' work environments and work histories [30].

RESULTS

All IPCs reported that it was common for medical tourists to bring at least one friend or family member abroad with them unless they were specifically discouraged from traveling with a companion[a]. Family members, especially spouses, were the most common type of informal caregivers present at the facilities where participants worked. Facilities that offered in-patient procedures often had cots or beds available in patients' rooms for these informal caregivers. This arrangement demonstrates the intense physical proximity of this transnational caregiving practice. Figure 1 provides an example of this kind of patient room. In the case of out-patient clinics or where co-habitation in the patient's hospital room was uncomfortable or impossible, friends and family typically stayed at nearby hotels or rented apartments.

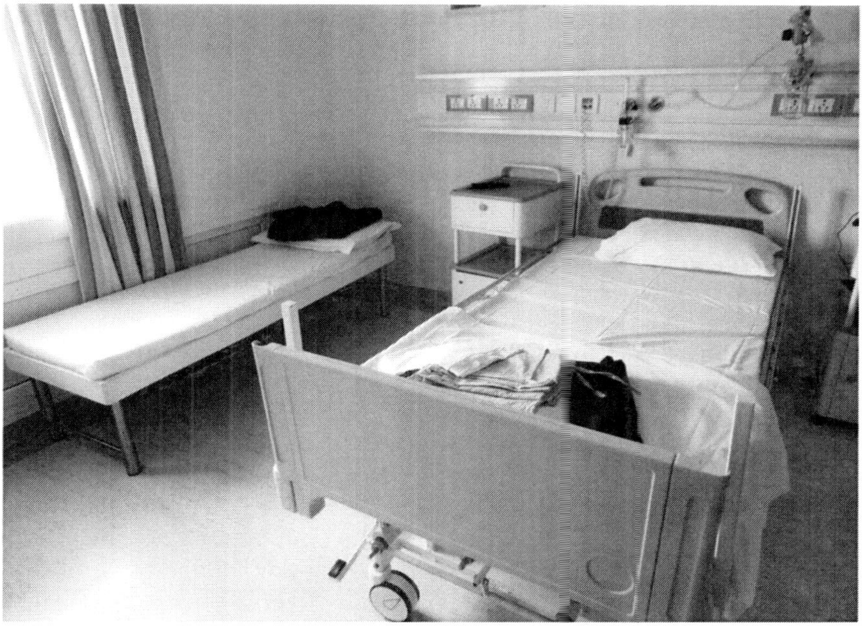

Figure 1. Typical medical tourist patient room. Taken at hospital in India that treats medical tourists, this photo conveys the close physical proximity that is experienced between some medical tourists and their informal caregivers while abroad. The cot on the left is for the friend or family member providing care while the bed on the right is for the patient. (Photo credit: authors).

Thematic analysis of interviews conducted with 21 IPCs about their interactions with and observations of medical tourists' informal caregivers revealed three roles commonly adopted by this group: (1) knowledge broker, (2) companion, and (3) navigator. The knowledge broker role refers to key functions undertaken by caregivers around the transfer of information between the medical tourist and facility staff or others. The companion role refers to functions around the provision of emotional and physical comfort or support as well as hands-on care. Finally, the navigator role encompasses functions dealing with information gathering and care coordination responsibilities assumed by medical tourists' informal caregivers. While IPCs observed that many caregivers adopt all three roles, often simultaneously, we consider them separately in this section in order to provide detailed accounts of the distinct features of each role. We include verbatim quotations throughout in order to 'give voice' to the participants. Each quotation is followed by a country name and a number. The name refers to the country in which the participant worked at the time of the interview while the number indicates the number of years they had worked as an IPC at that particular facility.

Knowledge broker

Participants emphasized that a key role caregivers play is that of knowledge broker. The facilitation of knowledge transfer from medical tourism facility staff to the patient by caregivers is typically done in four ways: inquiry, clarification, translation, and retention. First, caregivers commonly make inquiries on behalf of patients. Questions directed toward IPCs tend to be about logistics or planning, while those directed at formal providers cover clinical concerns such as prescriptions, care options, and advice for after discharge. Second, by seeking clarification caregivers help to ensure that patients clearly understand medical information. This activity can be especially helpful when patient rights are explained, when complications occur, or when patients are confused about what they have been told. Third, caregivers help with translation or basic communication when they are more proficient than the patient in the language or regional vernacular used by facility staff. "*When the patient comes over here, the language may differ or the accent may differ, [the understanding of] English between the patient and the [facility] staff may differ slightly, so there needs to be someone who can patiently communicate with the [facility] staff [if the patient cannot]*" (India, 2.25). Finally, caregivers retain information that has been conveyed by facility staff to the patients. "*Sometimes when you're in that position [as a patient], you're ill, you're in a bed, the doctor comes in and says x, y or z, you don't remember it... So it's nice to have someone there with you who will be able to retain that information*" (Barbados, 2). Knowledge brokering by facilitating the transfer of information from facility staff to patients can be demonstrative of collaboration between caregivers and formal providers and other facility staff.

Caregivers often engage in knowledge brokering by offering unprompted but useful information about medical tourists to facility staff. IPCs noted that caregivers frequently offer this kind of information while patients are at the facility and also following discharge. For example, caregivers may voice patients' concerns or articulate complaints when patients are reluctant to make such remarks. They may also help formal providers obtain correct information when patients are untruthful (e.g., purposely not reporting their correct weight), do not accurately recall their health history, or are unwilling to communicate. For example, one participant reported that she once worked with a patient who was uncommunicative and therefore the accompanying caregivers answered her questions instead: *"...sometimes the patient didn't feel like talking and then his mother or his wife would give me the information"* (Mexico, 3). These types of information exchanges between facility staff and caregivers, ones that are not prompted by the medical tourist and may not occur in the patient's presence, reveal *"...the good, the bad, the ugly, everything [about the patient that may be helpful to facility staff]"* (Costa Rica, 2.5). In other words, caregivers may broker information or share details with facility staff to ensure that they have accurate information with which to make decisions that affect patients' health independent of requests to do so from medical tourists.

Participants observed that caregivers often act as information liaisons between patients and friends and family members back home. They provide friends and family members with updates by phone, email, Skype, and even letters. One participant recounted that caregivers commonly *"bring their computer [or] they can borrow our telephones and right after the surgery they...communicate with people from home just to let them know that everything went okay"* (Mexico, 3). Acknowledging the importance of this aspect of caregivers' roles, some of facilities even offer them free international calling. This dimension of knowledge brokering serves to underscore the truly transnational nature of the roles that caregivers take towards medical tourists as well as within the larger practice of medical tourism.

Companion

One way that informal caregivers engage in the companion role is by creating a feeling of emotional safety and security for the medical tourist. A participant reported that caregivers provide *"... emotional support, which was probably the most crucial [thing that they can do]. It's something that a medical tourism company cannot offer"* (Turkey, 1.5). Caregivers are familiar to patients, they have established relationships with one another, and bonds of trust already exist before their journeys begin. These traits put them in a unique position to offer better quality support, including moral and spiritual, relative to facility staff. Caregivers' trusting relationships with medical tourists often makes them aware of patients' preferences and needs. All the IPCs with whom we spoke indicated that trusting relationships between patient and caregiver can have emotional benefits for the medical tourist and that this social bond is a significant reason why they encourage medical tourists to avoid travelling alone. To a medical tourist, a caregiver is *"someone there who would be able to share your...experience...you wouldn't have to...be looking for words, you would just be able...to flow with this person because this is someone that you know, it's someone you have a history with, so you'd be more comfortable"* (Barbados, 2). According to the participants, the presence of the caregiver can be beneficial to creating a feeling of safety and security for the patient.

Many participants noted that caregivers typically try to address medical tourists' comfort in their role as companions. This assistance comes in the form of: providing the patient with wanted or needed items, monitoring symptoms, and helping the patient deal with the 'foreignness' of the destination country. IPCs commonly observed caregivers obtaining items for medical tourists inside and outside the facility because, participants postulated, patients were more comfortable asking their companions than their IPC. For example, patients who dislike the food served at the facility can send caregivers into the surrounding area on regular food purchasing trips. To ensure physical comfort, nearly all participants noted that caregivers monitor patients' symptoms and alert formal providers of noteworthy changes. A participant explained: *"...we provide them with our extension numbers and they might call us just to say '...my wife is having nausea after the operation, do you think somebody can give her some medicine'. So we do quite often see that situation"* (Thailand, 6). Caregivers may further ensure patients' comfort by being a familiar, reliable figure in an otherwise unfamiliar environment. IPCs explained that being outside one's own country can be inherently stressful. Language differences in particular can be especially difficult for the patient. Patients and caregivers *"really just have each other... I mean the outside world speaks a different language, different culture. They rely a lot on each other"* (Turkey, 1.5).

Many participants reported that caregivers commonly provide hands-on care, especially after the patient is discharged from the facility. This care may take the form of assistance with mobility and everyday tasks and help with following clinical advice. Patients' mobility and their capacity to perform everyday tasks after surgery are typically minimal, and therefore some facilities require that medical tourists bring a caregiver to assist them with daily activities following the operation. Caregivers often help patients with dressing, showering, toileting, and mobility at the facility and after discharge. After discharge "*... patients...might need the help of a companion with luggage, with getting in and out of a vehicle, perhaps getting up to their hotel room, go out to dinner, things like that*" (USA, 8). Caregivers also help patients follow clinical advice such as taking medications, sometimes offering hands-on care to ensure this advice is followed. One participant explained that caregivers need to know "*what not to do [when caring for the patient]: do not...bend the knee over this position if he has a knee replacement, or do not give him anything to eat besides what the doctor says if he has a gastric sleeve...*" (Costa Rica, 0.6). In some cases, caregivers continue this aspect of their companionship role after returning to their home country, ensuring appropriate post-operative and follow-up care regimens are followed.

Navigator

Caregivers fill a navigator role when they guide the patient through various aspects of the medical tourism experience. One such aspect is geographical and cultural navigation. For example, caregivers typically gather tourist information. As mentioned above, they may do this by asking facility staff, especially IPCs, questions about particular destinations in the local area. Some of these questions may pertain to tourist activities that are suitable for the patient after discharge. Caregivers may also seek location-specific information from IPCs, such as "*how to take a taxi and...what areas are safe to go to and what aren't...*" (Mexico, 12). In addition to gaining familiarity with navigating the destination country or city, caregivers typically learn how to navigate the destination facility and transmit this information to the medical tourist. According to one participant, caregivers are "*the ones who read through the instructions [about the facility] and enforce them [with the patient]...*" (Mexico, 1.5, 0.6). Participants also noted that, in their capacities as navigators, all caregivers familiarize themselves with useful locations in the facility such as bathrooms and magazine vendors.

Most participants observed that a large part of a caregiver's navigator role involves coordinating paperwork and gathering required documentation. "*Usually the companion, to relieve the patient that's having the surgery, does all the running around to make it happen*" (Costa Rica, 5). Many IPCs reported that completing paperwork is one of the first tasks that caregivers must undertake after arriving at the facility. This task requires completing forms for the patient and verifying that patient information is accurate. They also "*...tend to want to deal with the finances...they're always very worried about [the patient] being worried about...the balances or the costs that their stay incur. They tend* *to want to shield [the patient] from that*" (Barbados, 2). They typically monitor and complete financial paperwork and may access a bank in person or electronically to exchange currency, to ensure funds are available, and to get money for airfare, treatment, and other expenses. Caregivers also commonly transfer documentation from the facility abroad to patients' regular physicians and vice versa before departure and upon return home. These documents include medical records, letters, prescriptions, and/or test results. IPCs remarked that navigating the coordination of paperwork and documentation is generally done by caregivers to minimize the number of concerns or stressors that medical tourists encounter.

Creating plans and ensuring that travel itineraries are followed are both common responsibilities for caregivers in their roles as navigators. Caregivers sometimes play a large role in preparing for the trip abroad. They communicate with IPCs to organize the trip, sometimes as early as sending the initial inquiry. They continue to fill this role throughout the stay by arranging lodging, tours, and food options. They may further help patients by arranging ground transportation and international flight arrangements. "*...They get the wheelchair for them and, you know rent a chair car to transport the patient to and from the hotel...*" (USA, 8). Caregivers also sometimes create itineraries with the help of IPCs, and typically ensure that these plans are followed by the medical tourists. A key reason for this is because medical tourists typically look to caregivers "*for [information about] the whole journey, you know: 'What time are we going to the doctor? What time is our appointment? What time is the check-up? When do we go next? Tomorrow's a day off, are we going touring? Are we going to the beach? Are we going shopping?'*" (Israel, 3). Caregivers know the estimated length of the patient's stay in the facility and note changes in the recovery schedule after surgery so that they can alter or cancel activities as needed. Even after they are back in their home country, caregivers often help patients create and follow an itinerary of follow-up appointments. The majority of participants noticed that caregivers generally help patients navigate their experience abroad by making and altering plans as needed, and that nearly all of them play an important role keeping patients on schedule.

DISCUSSION

The analysis has found that medical tourists' informal caregivers adopt three major roles that complement those of formal health care providers: knowledge broker, companion, and navigator. As knowledge brokers, they facilitate the transfer of information between the medical tourist and the formal health care providers, the IPC, friends and family members back home, and others. As companions they support the medical tourist emotionally by ensuring their comfort and providing hands-on care. Finally, in their role as navigators they gain familiarity with the destination country, coordinate the trip, and handle facility documents. In this section, we examine interconnections between these roles and consider all three in light of the existing medical tourism and informal caregiving literatures. We also identify future directions for research emerging from the findings.

Each of the three caregiver roles identified by IPCs encompass several responsibilities. Many responsibilities are unique to a single role. Handling finances, for example, which is part of the navigation role, has no overlap with other role components, though the outcomes of this responsibility might have implications for caregivers' other roles. Other responsibilities might seem less clearly delineated with regard to the roles to which they contribute because of the similarities of the activities they involve. For example, there are information exchange aspects to all the roles. The difference is the intermediary or primary position of the caregiver in the exchange and the nature of the information. Offering or retaining information on behalf of the patient is characteristic of the knowledge broker role. Meanwhile, flagging the patient's changing health status to staff relates most closely to the companion role, whereas discussion about travel logistics relates to the navigator role. By both focusing on the intent of an action being undertaken and having identified the scope of each role it was clear in the analytic process as to which responsibilities were attributed to what roles.

A strong relationship exists between the caregiver roles because the actions, activities, and overall responsibilities undertaken in one role can have implications for the other two roles. In this way, there is overlap between all the roles identified in the findings. By helping patients follow clinical advice, which pertains to the companion role, caregivers act on advice they might have retained in their capacity as a knowledge broker. When the clinical advice requires arranging medical appointments and ensuring that patients arrive at appointments in a punctual manner, then there is also overlap with the navigator role. Knowledge brokering and the companion role further overlap when caregivers monitor medical tourists' symptoms and communicate those observations to a health care provider; symptom monitoring is identified as part of the companion role, whereas voicing comments or concerns about symptoms to formal providers is part of the knowledge broker role. These are but a few of the many examples of the ways that distinctive roles become interconnected through the practice of informal caregiving in medical tourism. Although these interconnections can create some overlaps or redundancies between roles, we believe that the distinctions between the knowledge broker, companion, and navigator roles remains useful for clearly positioning the friends and family who accompany medical tourists abroad as informal caregivers, and ultimately unpaid health care providers, within the industry.

Whereas the existing medical tourism literature does not discuss informal caregivers' roles in detail, there are some mentions of the same activities or responsibilities discussed by the IPCs with whom we spoke. Solomon [31] peripherally mentions caregivers' research responsibilities associated with the initial inquiry as well as the fact that they have in-facility information seeking interactions with IPCs. These findings from Solomon's ethnographic study, though not a central part of his analysis, parallel some of our own research findings. Kangas' [32] ethnographic research into the travel of Yemeni patients to other countries for private medical care confirms the centrality of family members in making the initial decision to access care abroad and their knowledge brokering roles. Kingsbury et al. [33] state that caregivers sometimes need to assume essential decision-making responsibilities and that unanticipated changes in medical tourists' health status can force them to "navigate shifting boundaries of...[their] roles." Two autobiographical narratives written by caregivers about their journeys abroad with medical tourists offer the strongest source of confirmation of the findings reported herein [27, 28]. The authors of both narratives disclosed their active participation in all three of the roles identified here. Indeed, these former informal caregivers engaged in most of the responsibilities attributed to each of the knowledge broker, companion, and navigator roles (see [33] for greater analytic discussion of these narratives). With the findings of this article, we assist with putting discussion of caregiver activities and responsibilities from these other sources into context by explicitly considering the roles to which they are contributing.

The existing informal caregiving literature demonstrates that there are indeed commonalities between the roles filled by medical tourists' caregivers and those adopted by other types of informal caregivers. Much of this literature is focused on the practice of informal caregiving in the home (e.g., [6, 8, 10, 25, 34]). One difference between these caregivers and those discussed in this article is the person who serves as a first point of contact and ongoing informational resource. It is the IPC for medical tourists' caregivers, whereas homecare nurses often serve in this capacity for 'conventional' (i.e., local, non-transnational, long-term) informal caregivers in the home [35, 36]. Furthermore, our findings reveal that medical tourists' caregivers sometimes arrange lodging and tours and transfer documents to and from the destination country, whereas these are not common activities for non-transnational caregivers [8, 34, 37]. Differences such as these can be attributed to the specific context of care (i.e., local vs. transnational care and familiar vs. unfamiliar environment). The greatest overlaps between roles undertaken by medical tourists' caregivers and other types of caregivers come in relation to non-context-dependent functions, such as facilitating information transfer and engaging in symptom monitoring. Many studies of informal caregiving in the home have found that informal caregivers in this setting play a pivotal role in transmitting details on changes in patients' health status to formal health care providers, keeping logs of symptom changes, and other similar activities [11, 37, 38, 39]. There are also cases in which the specific activities undertaken by medical tourists' caregivers are different from those of conventional informal caregivers but the overall roles being enacted are identical. For example, finding a wheelchair to aid in a medical tourist's mobility in an airport is likely done with the same intent as arranging for a patient's accessible transit pick-up at home, a common activity undertaken by caregivers in other contexts [34, 37]; in both cases the caregiver functions as a navigator.

Limitations

This analysis has some limitations. First, using semi-structured phone interviews to obtain data has some intrinsic limitations. The linguistic diversity of participants, for example, was limited. The primary interviewer's ability to speak only two languages with fluency, French and English, disqualified potential participants who were not fluent in either, although one Spanish-language interview was accommodated. Furthermore, unspoken insights that may have been gleaned through the interviews were missed because they were not conducted face-to-face. The reliability and affordability of phone and Skype interviews (see

[40]) outweighed these concerns as conducting interviews using this medium is what enabled us to recruit such an international sample of participants.

Second, although interviewing IPCs is useful, it results in a second-hand perspective on medical tourists' informal caregivers. We believe that the strengths of interviewing this group in the context of this study, particularly that they were able to comment on trends across many caregivers, outweighed this limitation. Every interview we conducted captured experiences with hundreds of caregivers, which was necessary to create the broad understanding that we sought to provide with this analysis. However, our analysis needs to be complemented with first-hand accounts from caregivers, and we intend on pursuing this in the next phases of our research. We believe that the limitations we have identified here are acceptable and do not significantly compromise the rigor of the study or analysis.

Future research directions

Although we provide the first dedicated investigation of the roles of informal caregivers in medical tourism, this article contributes only a small part to our understanding of them. We have not measured the frequency with which each role is adopted, we have not assessed the spatiality or temporality of these roles, nor have we considered the effectiveness of these roles in ameliorating or maintaining patient health or wellness or even offsetting what is known as 'caregiver burden'. These all serve as important directions for future research. For example, the collective impact of the stressors encountered by informal caregivers in the practice of care is referred to as caregiver burden [41, 42, 43]. Significant burden might lead to 'caregiver burnout' [41, 44, 45, 46]. In this article, we show that the friends and family members who accompany medical tourists abroad are indeed filling informal caregiving roles and it would be valuable to determine if and how the activities they undertake in their capacities as knowledge brokers, companions, and navigators result in exposure to burden and ultimately burnout. This knowledge could then be used to assist in identifying interventions to offset that burden if it is found to exist. The caregiver group could benefit from, for example, the development of informational tools that anticipate and address the possibility of caregiver burnout. Although there has been some discussion of the need to create reliable, evidence-informed informational tools about medical tourism and enhance patients' access to such sources [47, 48], there has been no consideration of whether or not medical tourists' informal caregivers could similarly benefit from enhanced access to credible information. Second, any demonstration of burden or burnout among this group would contribute new evidence to the health equity debates that exist around the practice of medical tourism that focus on discerning who 'benefits' and who 'loses' from the existence of this health services trade (see, for example, [21, 49]). Although informal caregivers are not addressed in contemporary analyses of the health equity effects of medical tourism, they need to be incorporated into examinations of this subject.

An important area for future research pertains to our participant group: IPCs. An enhanced understanding of the scope and scale of IPCs' roles and responsibilities as a whole is very much needed. Such knowledge is critical given that many aspects of informal caregivers' roles hinge on the time, energy, and attentiveness of IPCs. Yet, there is not an adequate understanding of IPC training regarding interactions with caregivers, or the proportion of their time allocations or work tasks assigned to dealing with such interactions. Furthermore, we have shown that IPC and informal caregiver roles intersect at many points in a medical tourist's journey. The implications of these intersections for the roles assumed by each group are unknown, such as the benefits and drawbacks for the health and wellbeing of the medical tourist. Such research would not only be important in increasing the knowledge of caregiver roles and responsibilities, it would also expand the existing knowledge of relationships between other medical tourism stakeholder groups. Dedicated attention to IPC roles and responsibilities could, for example, be used to create a schematic that demonstrates the intersection of the roles of IPCs, formal health care providers, other staff from the medical facility and from hospitality services such as hotel concierges and airline representatives, caregivers, and medical tourists.

CONCLUSIONS

We found that the friends and family members who accompany medical tourists abroad engage in caregiving roles almost continuously while abroad. They facilitate and supplement the efforts of formal health care providers at medical tourism facilities to ensure the patient's health and wellbeing and shoulder some of the responsibilities that might otherwise be assigned to patients. In this way, caregivers can act as both amplifiers and buffers: they amplify the efforts of the facility staff through facilitation and supplementation while buffering medical tourists from stresses stemming from responsibilities, anxieties, and discomforts. Although there are parallels between some of our findings about caregiver roles and those shared in the existing caregiving literature, the unique transnational care context of informal caregiving in medical tourism reveals activities and responsibilities assigned to particular roles that are specific to this particular care practice.

We believe that the knowledge that has been gleaned about a heretofore neglected medical tourism stakeholder group, namely patients' informal caregivers, and the roles filled by members of this group provides valuable insight into how the medical tourism industry operates. Given the integral roles that friends and family members play (and particularly while they are abroad), the practice of medical tourism and thus the industry that supports it seems highly dependent on their unpaid care work. As with other informal caregivers, these individuals are effectively overlooked "shadow workers" [50] – unpaid, untrained, and largely unrecognized care providers - in what is often reported as a highly lucrative industry. We believe that adopting this critical perspective in light of the findings is essential to effectively address the question 'who ultimately benefits from medical tourism?' that is central to the health equity debates that surround the medical tourism industry. Although this question is far beyond the scope of this article, we do demonstrate the value of considering the roles that *every* stakeholder group plays in enabling this global health services practice. Without having such knowledge it becomes impossible to effectively determine the full scope of the health equity impacts of medical tourism, mitigate the negative ones, and enhance the positive ones, which is a key issue currently being discussed in research and policy circles [6, 14,21, 51, 52].

ENDNOTE

[a]Informal caregivers to patients recovering from cosmetic or bariatric surgeries were sometimes noted to be problematic to patients' wellbeing because they can become visibly distressed by a patient's appearance. Therefore, some IPCs who worked at facilities specializing in these procedures advised medical tourists to travel unaccompanied. See [53] for further examination.

ACKNOWLEDGEMENTS

We would like to thank those who participated in this study for sharing their experiences and views with us. Thank you to Alejandro Ceron for conducting the Spanish language interview. The study was funded by an Operating Grant from the Canadian Institutes of Health Research (funding reference #115692). VAC is also funded by a Scholar Award from the Michael Smith Foundation for Health Research.

COMPETING INTERESTS

The authors declared no potential conflicts of interest with respect to the research, authorship, and/or publication of this article.

AUTHORS' CONTRIBUTIONS

VC conducted the interviews, led the coding, and played a leadership role in writing this article. VAC is lead investigator on the study that funded this research and developed the protocol. She contributed to all aspects of the study and was heavily involved in preparing this manuscript. JS and LT contributed input to the data collection instrument, participated actively in the face-to-face analysis meeting, and provided feedback on drafts of this paper. JS also aided in recruitment. All authors reviewed and approved the submitted manuscript.

REFERENCES

1. Martin R: Clark Howard Discusses His Cancer Diagnosis With Dr. Randy Martin. 2009, Atlanta: WSB-TV,http://www.wsbtv.com/news/news/clark-howard-discusses-his-cancer-diagnosis-with-d/nFBhr/,

2. Smeltzer SC, Bare BG, Hinkle JL, Cheever KH: Health care delivery and nursing practice. Brunner and Suddarth's Textbook of Medical-Surgical Nursing: In One Volume. 2010, Philadelphia: Lippincott Williams & Wilkins, 9-10. 12

3. Schein C, Gagnon AJ, Chan L: The association between specific nurse case management interventions and elder health. J Am Geriatr Soc. 2005, 53: 597-602.

4. Schoenborn NL, Arbaje AI, Eubank KJ, Maynor K, Carrese JA: Clinician roles and responsibilities during care transitions of older adults. J Am Geriatr Soc. 2013, 61: 231-236.

5. GG Van Bruchem-van de S, van der Arend AJG, Huijer Abu-Saad H, Spreeuwenberg C, van Wijmen FCB, ter Meulen RHJ: The role of nurses in euthanasia and physician-assisted suicide in The Netherlands. J Med Ethics. 2008, 34: 254-258.

6. Connor S, Adams J: Caregiving at the end of life. Hastings Cent Rep. 2003, 32 (Suppl 2): S8-S12.

7. Karimli L, Ssewamala FM, Ismayilova L: Extended families and perceived caregiver support to AIDS orphans in Rakai district of Uganda. Child Youth Serv Rev. 2012, 34: 1351-1358.

8. Ussher JM, Wong WKT, Perz J: A qualitative analysis of changes in relationship dynamics and roles between people with cancer and their primary informal carer. Health. 2011, 15: 650-667.

9. Halvorsen K, Førde R, Nortvedt P: The principle of justice in patient priorities in the intensive care unit: the role of significant others. J Med Ethics. 2009, 35: 483-487.

10. Lam RC: Contradictions between traditional Chinese values and the actual performance: a study of the caregiving roles of the modern sandwich generation in Hong Kong. J Comp Fam Stud. 2006, 37: 299-313.

11. Sinding C: Informal care – two tiered care? the work of family members and friends in hospitals and cancer centres. J Sociol Soc Welfare. 2004, 31: 69-86.

12. Connell J: Contemporary medical tourism: conceptualisation, culture and commodification. Tourism Manage. 2013, 34: 1-13.

13. Ehrbeck T, Guevara C, Mango PD, Cordina R, Singhal S: Health care and the consumer. McKinsey Quart. 2008, 4: 80-91.

14. Lunt N, Smith R, Exworthy M, Green ST, Horsfall D, Mannion R: Medical tourism: treatments, markets and health system

implications: a scoping review. 2012, Geneva OECD Directorate for Employment, Labour and Social Affairs,http://www.oecd.org/els/health-systems/48723982.pdf.

15. Deloitte Center for Health Solutions: Medical tourism consumers in search of value. 2008, New York: Deloitte, http://www.deloitte.com/assets/Dcom-unitedStates/Local%20 Assets/Documents/us_chs_MedicalTourismStudy(3).pdf

16. NaRanong A, NaRanong V: The effects of medical tourism: Thailand's experience. Bull World Health Organ. 2011, 89: 336-344.

17. Pocock NS, Phua KH: Medical tourism and policy implications for health systems: a conceptual framework from a comparative study of Thailand, Singapore and Malaysia. Global Health. 2011, 7: 12-

18. Tourism Research and Marketing: Medical tourism: a global analysis. 2006, Arnhem: Association for Tourism and Leisure Education,http://www.atlas-webshop.org/Medical-tourism-a-global-analysis,

19. Alleman BW, Luger T, Schacht Reisinger H, Martin R, Horowitz MD, Cram P: Medical tourism services available to residents of the United States. J Gen Intern Med. 2011, 26: 492-497.

20. Burkett L: Medical tourism: concerns, benefits, and the American legal perspective. J Legal Med. 2007, 28: 223-245.

21. Johnston R, Crooks VA, Snyder J, Kingsbury P: What is known about the effects of medical tourism in destination and departure countries? a scoping review. Int J Equity Health. 2010, 9: 24.

22. Kangas B: Hope from abroad in the international medical travel of Yemeni patients. Anthropol Med. 2007, 14: 293-305.

23. Yeoh E, Othman K, Ahmad H: Tourism management understanding medical tourists: word-of-mouth and viral marketing as potent marketing tools. Tourism Manage. 2013, 34: 196-201.

24. Yu JY, Ko TG: A cross-cultural study of perceptions of medical tourism among Chinese, Japanese and Korean tourists in Korea. Tourism Manage. 2012, 33: 80-88.

25. Bevan JL, Pecchioni LL: Understanding the impact of family caregiver cancer literacy on patient health outcomes. Patient Educ Couns. 2008, 71: 356-364.

26. Molloy GJ, Johnston DW, Witham MD: Family caregiving and congestive heart failure: review and analysis. Eur J Heart Fail. 2005, 7: 592-603.

27. Grace M: State of the Heart: A Medical Tourist's True Story of Lifesaving Surgery in India. 2007, Oakland: New Harbinger Publications

28. Rose DA: Larry's Kidney: Being the True Story of How I Found Myself in China with My Black Sheep Cousin and His Mail-Order Bride, Skirting the Law to Get Him a Transplant–and Save His Life. 2009, New York: William Morrow

29. Fereday J, Muir-Cochrane E: Demonstrating rigor using thematic analysis: a hybrid approach of inductive and deductive coding and theme development. Int J Qual Method. 2006, 5: 80-92.

30. Braun V, Clarke V: Using thematic analysis in psychology. Qual Res Psychol. 2006, 3: 77-101.

31. Solomon H: Affective journeys: the emotional structuring of medical tourism in India. Anthropol Med. 2011, 18: 105-118.

32. Kangas B: The burden of pursuing treatment abroad: three stories of medical travelers in Yemen. Global Soc Pol. 2010, 10: 306-314.

33. Kingsbury P, Crooks VA, Snyder J, Johnston R, Adams K: Narratives of emotion and anxiety in medical tourism: on state of the heart and Larry's kidney. Soc Cult Geogr. 2012, 13: 361-378.

34. National Alliance for Caregiving, AARP: Caregiving in the U.S. 2004, Bethesda, Washington: National Alliance for Caregiving, AARP,http://www.caregiving.org/data/04finalreport.pdf,

35. Mezey M: Nurses and their changing relationship to family caregivers. The Cultures of Caregiving: Conflict and Common Ground Among Families, Health Professionals, and Policy Maker. Edited by: Levine C, Murray TH. 2004, Baltimore: The John Hopkins University Press, 70-86.

36. Neufeld A, Harrison M: Nursing and Family Caregiving: Social Support and Nonsupport. 2010, New York: Springer Publishing Company

37. Donelan K, Hill CA, Hoffman C, Scoles K, Hollander Feldman P, Levine C, Gould D: Challenged to care: informal caregivers in a changing health system. Health Affair. 2002, 21: 222-231.

38. Petronio S, Sargent J, Andea L, Reganis P, Cichocki D: Family and friends as healthcare advocates: dilemmas of confidentiality and privacy. J Soc Pers Relat. 2004, 21: 33-52.

39. Weinberg DB, Lusenhop RW, Hoffer Gittell J, Kautz CM: Coordination between formal providers and informal caregivers. Health Care Manage Rev. 2007, 32: 140-149.

40. Laws S, Harper C, Marcus R: Telephone interviews. Research for Development. 2003, Thousand Oaks, CA: SAGE, 298-299.

41. Lee H, Singh J: Appraisals, burnout and outcomes in informal caregiving. Asian Nurs Res. 2010, 4: 32-44.

42. Ray RA, Street AF: The dynamics of socio-connective trust within support networks accessed by informal caregivers. Health. 2011, 15: 137-152.

43. Roth DL, Perkins M, Wadley VG, Temple EM, Haley WE: Family caregiving and emotional strain: associations with quality of life in a large national sample of middle-aged and older adults. Qual Life Res. 2009, 18: 679-688.

44. Murphy NA, Christian B, Caplin DA, Young PC: The health of caregivers for children with disabilities: caregiver perspective. Child Care Health Dev. 2007, 33: 180-187.

45. Takai M, Takahashi M, Iwamitsu Y, Oishi S, Miyaoka H: Subjective experiences of family caregivers of patients with dementia as predictive factors of quality of life. Psychogeriatr. 2011, 11: 98-104.

46. Yılmaz A, Turan E, Gundogar D: Predictors of burnout in the family caregivers of Alzheimer's disease: evidence from Turkey. Australas J Age. 2009, 28: 16-21.

47. Lunt N, Hardey M, Mannion R: Nip, tuck and click: medical tourism and the emergence of web-based health information. Med Inform J. 2010, 4: 1-11.

48. Penney K, Snyder J, Crooks VA, Johnston R: Risk communication and informed consent in the medical tourism industry: a thematic content analysis of Canadian broker websites. BMC Med Ethics. 2011, 12: 17-

49. Chen YYB, Flood CM: Medical tourism's impact on health care equity and access in low- and middle-income countries: making the case for regulation. J Law Med Ethics. 2013, 41: 286-300.

50. Armstrong P, Armstrong H, Coburn D: Unhealthy Times: Political Economy Perspectives on Health and Care. 2001, London: Oxford University Press

51. Forgione DA, Smith PC: Medical tourism and its impact on the US health care system. J Health Care Financ. 2007, 34: 27-35.

52. Hopkins L, Labonté R, Runnels V, Packer C: Medical tourism today: what is the state of existing knowledge?. J Public Health Pol. 2010, 31: 185-198.

53. Casey V, Crooks VA, Snyder J, Turner L: "You're dealing with an emotionally charged individual...": an industry perspective on challenges posed by medical tourists' informal caregiver-companions. Global Health. 2013, 9: 31.

9

CHAPTER

Promoting Social Responsibility Amongst Health Care Users: Medical Tourists' Perspectives on an Information Sheet Regarding Ethical Concerns in Medical Tourism

Krystyna Adams,[1] Jeremy Snyder,[1] Valorie A Crooks,[2] and Rory Johnston[2]

[1]*Faculty of Health Sciences, Simon Fraser University, Blusson Hall, 8888 , University Drive, Burnaby, British Columbia, Canada*

ABSTRACT

Background

Medical tourism, the practice of patients traveling out of country with the intention to receive medical care paid for out of pocket, is considered an expanding industry globally [1,2]. While there is a dearth of empirical evidence on patient flows and the impacts of medical tourism, media coverage often provides anecdotal stories that indicate the potential for surgical complications, and poor quality or unnecessary care due to a lack of regulation in the industry [3]. Partially due to the lack of empirical evidence available about medical tourism, researchers are concerned about the gaps in the information medical tourists may be accessing regarding potential impacts of medical tourism on both individual safety and health equity [4,5].

Medical tourism is receiving increasing attention, both as a developing industry and a topic of research interest [6]. While there appears to be a wide variety of medical tourism flows, including travel between both countries of similar and different levels of economic development, information available to medical tourists demonstrates a concerted effort on behalf of low and middle income countries to entice persons to travel from developed to developing countries for medical care [7,8]. This recent trend of travel from more developed to developing countries has become a focus for research on the impacts of medical tourism on global health equity [9]. Discussions by researchers about growing health inequities globally have identified the growth of the medical tourism industry as potentially one cause of these inequities [10]. The rise of neoliberalism and economic globalization in the past few decades has resulted in health care being increasingly conceptualized as a tradable commodity across borders. The medical tourism industry markets to patients as consumers of health care, encouraging individuals to take responsibility for their health by taking advantage of potential cost savings of medical treatment in developing countries and high quality care available in private facilities catering to medical tourists [11]. Patients may be motivated to travel to access procedures which are domestically unavailable or require undesired waiting times, receive care and recover in a relaxing or exotic landscape, and/or reduce costs for procedures domestically paid for out-of-pocket [7].

Research indicates that many Canadian medical tourists use word-of-mouth or the internet to inform their decisions regarding this practice, which may result in decision-making that is uninformed or biased around individual safety and treatment outcomes [12]. Canadian medical tourists may also be unaware of ethical concerns regarding potential impacts of medical tourism on global health equity, particularly potential impacts to destination and departure country health resources and health care systems [13,14]. While researchers are increasingly discussing these potential ethical concerns related to medical tourism, medical tourists may be unable to act in a socially responsible manner if uninformed of these impacts [15].

Here we explore the responses of Canadian medical tourists to an information tool intended to encourage more informed and ethical decision-making around medical tourism. This tool was developed through an iterative process that drew on values from guidelines in related domains as well as public health communication research to effectively inform Canadians about ethical concerns related to medical tourism. Through analyzing feedback received in one-on-one interviews with former Canadian medical tourists, in this paper we examine the potential usefulness and impacts of such a tool on patients' decision-making, as well as the contexts in which such a tool might better inform and promote social responsibility amongst Canadian medical tourists and ultimately a more ethical medical tourism. Furthermore, this research provides insight into considerations for the field of public health in promoting social responsibility in the provision and utilization of health care.

Ethical issues in medical tourism and the need for enhanced awareness

Despite a lack of empirical evidence on the impacts of medical tourism, researchers have indicated several ethical concerns relating to medical tourism's potential impact on both an individual and societal level within destination and departure countries, as well as on a global scale [13,14]. Medical tourism may negatively impact upon individual health due to a lack of risk communication to patients and/or inadequate informed consent [2,14]. Furthermore, the medical tourism industry may divert resources from the public to private sector. Increasing privatization of health care may reorient priorities in the provision of health care to provide more profitable specialized care, including care which meets the needs of foreign patients travelling for some form of specialized care [14,16]. This reorientation of health care may reduce access to desired health care for local patients in medical tourism destination communities [1]. In countries of origin for medical tourists, potential complications affiliated with procedures obtained abroad, particularly due to a lack of regulation in the industry, may divert resources to treat these complications. Furthermore, reduced pressure for system reform if patients leave the health care system to seek out care abroad may result in only those persons able to travel as a medical tourist accessing health care [1,13]. Finally, increased movement of patients across borders to access medical care may increase the transmission of infectious diseases [17]. Concerns for medical tourism's negative impacts on health outcomes and health equity indicate a burden of responsibility for those engaging in medical tourism to consider these impacts and act in a socially responsible manner when providing and utilizing health care [14].

The primary ethical concerns surrounding the development of medical tourism and its impacts are also related to larger impacts on global health equity. While countries that are promoting themselves as medical tourism destinations cite economic motivations for industry growth [18], these economic advantages may come at the cost of undermining initiatives that intend to improve health equity both within and between nations. Increases in privatized health care globally have detracted from health care system strengthening required to meet the goals of "Health for All" outlined in the World Health Organization's Declaration of Alma Ata [19] and protect the human right to health [20]. While some medical tourism stakeholders have mentioned potential benefits of the practice to health outcomes and health equity, including improved quality control due to international accreditation of medical tourism facilities, increased training opportunities for health professionals treating foreign patients, and health worker retention in places experiencing high levels of brain drain to other countries, lack of regulation by some destination governments and the industry itself may neglect considerations for health outcomes and health equity in stakeholder decision-making [9].

Ethical concerns for medical tourists' destination and departure countries are related to the provision of health care within national borders. For destination countries, concerns about the growth of the medical tourism sector are related to the impacts of medical tourism on public health care due to the potential growth of the private sector [8]. The promotion of medical care to foreign patients may encourage a shift in health resource allocation resulting in a potential brain drain of health human resources from the public to the private sector and the diversion of public resources such as land or public finances to medical tourism businesses [21]. Medical tourism may incentivize training of health workers for more curative, costly and complicated procedures that are appealing to medical tourists [7]. This emphasis on more curative care may result in neglect for the provision of appropriate primary health care to local populations and an increasing normalization of privatized health care [1]. For departure countries, medical tourism raises concerns for these countries' health systems due to the potential diversion of resources to treat medical complications resulting from patients engaging in medical tourism [14]. Furthermore, medical tourism raises concerns for health equity within nations if only patients willing and able to travel for medical care are able to access certain medical procedures [1].

Ethical concerns regarding patient health in medical tourism are particularly focused on the lack of neutral information provided to medical tourists regarding risks and safety concerns and the implications of this for patients' abilities to achieve informed consent.

These safety concerns include potential medical complications related to the procedure itself, potential medical complications from travelling following a medical procedure, and/or inadequate continuity of care after returning home [17]. Patients may be unaware of these safety concerns due to lacking familiarity with the destination health care system, language or cultural barriers to adequate communication between medical tourists and health care workers, and/or lacking transparency regarding quality of medical facilities or health care workers [2].

Medical tourists and stakeholders in the medical tourism industry may be unaware of the ethical concerns outlined above, and as such, are unable to act in a socially responsible manner [14]. This lack of awareness may be due to biased sources of information, including sources with a vested interest in the profitability of the medical tourism industry [22]. Socially responsible engagement in the medical tourism industry requires individuals to act in a manner that respects the inherent dignity of all other humans [23]. Individuals are thus responsible for encouraging the development of social sustainability, defined as the creation of conditions that enable people to lead lives of personal value [24]. Improved health outcomes and health equity are increasingly recognized as cornerstones of social sustainability [25]. As such, the provision and utilization of health care contributes to social sustainability and necessitates responsible allocation and use of health resources in such a manner that optimizes health outcomes and health equity [26], even in relation to the private medical tourism industry. We contend that creating awareness of the ethical concerns of medical tourism will encourage socially responsible actions by individuals engaging in medical tourism, stakeholders in the medical tourism industry, and policy makers developing regulations for the industry [3]. It is for this reason that we have made an informational tool for Canadians considering medical tourism that prompts consideration of ethical, equity, and safety issues alike.

Informing individuals such as medical tourists about health issues can be done most effectively following extensive formative research to better understand the target audience and determine the ideal means of providing the information [27]. Thus, in the remainder of this article we present the findings of the formative, qualitative research we conducted in order to inform the development of an information tool for Canadians considering medical tourism that prompts consideration of ethical, equity, and safety issues. The process of formative research is useful for tailoring message points and determining effective dissemination of the information. Furthermore, formative research helps identify gatekeepers to the primary audience who either encourage or discourage information being provided to the primary audience [27]. Formative research activities such as focus groups and pre-testing of informational tools with primary target audiences and gatekeepers provide valuable insights that can be used for tool refinement and implementation [27,28]. With this in mind, in the sections that follow we examine feedback provided to us by intended users of the informational tool so as to assess the potential impact of such an information tool, inform revision and improvement upon the tool, and determine future steps for effective dissemination of the tool leading to more informed decision-making by Canadians considering medical tourism.

Methods

To develop the information tool on which the current formative study is based, an iterative process was utilized. This multi-step process is summarized in Figure 1 below. Here we report exclusively on the findings from the interviews with former Canadian medical tourists in Step 3 of this process. Although this tool was developed using ethical values and principles as a way to enhance consideration of ethical issues during the decision-making process, words such as 'ethics' and 'values' were deliberately excluded from the text as our previous research has shown that medical tourists can respond negatively to such terms, interpreting them to be explicitly judgmental. Instead, ethical concerns, along with equity and safety concerns were shared throughout the tool without being framed as such. As a health communication tool with the objective of informing and not guiding individual action, the target audience includes all Canadians that may be in the pre-contemplation or contemplation stage of decision-making regarding accessing care out of Canada as a medical tourist [29]. While this tool may provide useful information to all Canadians regarding this growing industry, it is intended to encourage more informed decision-making when considering or planning to travel out of country for private medical care, and as such, this stage of formative research focused on gaining feedback through semi-structured interviews with Canadians that have already travelled as medical tourists.

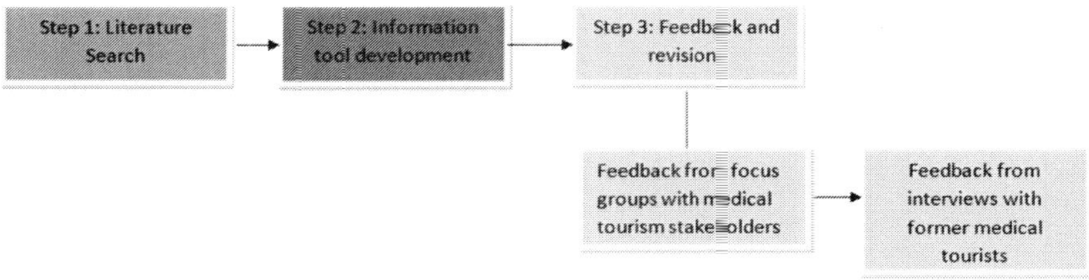

Figure 1. Multi-step process for information tool development.

Recruitment

Former medical tourists were selected to participate in this formative research as both a convenient sample to access in comparison to Canadians that may be considering participating in medical tourism, as well as a rich source of information regarding the impact of such a tool during the decision-making process given their experience. Participants were recruited using the following channels simultaneously: 1) Craigslist advertisements posted on sites for all major Canadian cities; 2) advertisement in a Vancouver newspaper; 3) posting invitations to participate in online medical tourism forums; and 4) snowball sampling through interview and focus group participants' networks. Information was provided for interested persons to contact either a toll-free phone number or email for further information and assessment of eligibility. Eligible potential participants were provided with a consent form to be read and signed before commencement of the interview. Participants were informed that prior to recruitment, the study was granted ethics approval from the Office of Research Ethics at Simon Fraser University.

Eligibility criteria for study required participants to be: 1) 18 years or older; 2) a holder of a provincial medical card; 3) someone who successfully pursued a surgical procedure outside of Canada that was neither a transplantation nor a reproductive surgery (as these procedures often involve third parties and raise distinct ethical issues); and 4) someone who paid privately for the surgery sought abroad and for whom the procedure was not performed based on a referral from a Canadian physician. Canadians who had travelled for transplantation or reproductive procedures were not included in this study due to the additional ethical concerns with these practices that are not covered in the information tool.

Medical Tourism & Me

Thinking about going abroad for surgery or other medical care? This is referred to as medical tourism. Here's some important information for you to consider before deciding whether or not to become a medical tourist.

My Health

1. *Did you know* that not all countries have the same hospital quality control standards we use here in Canada?
Before you decide whether to go, contact the Canadian Blood Services Agency to ask if they are aware of any blood supply or storage concerns in the country you are thinking about visiting.

2. *Did you know* that your medical record here in Canada should include your entire health history, even for medical care given in other countries?
Before you decide whether to go, ask the hospital you are planning to book your procedure with if they will give you records in English or French to bring home with you.

3. *Did you know* that diseases contracted abroad could be brought back home with you?
Before you decide whether to go, visit a travel clinic and make sure you get all recommended vaccines.

My Home

1. *Did you know* that some types of medical care require follow-up appointments or aftercare treatment?
Before you decide whether to go, make sure that you will be able to get the aftercare or follow-up appointments you need by talking to your regular doctor here about this.

2. *Did you know* that before you consent to any medical care you should know the risks of the procedure?
Before you decide whether to go, be sure to obtain information on procedure risks, ideally from a trusted source.

3. *Did you know* that your own decision to go abroad for medical care could actually impact other Canadians as well?
Before you decide whether to go, consider that if you experience medical complications you may need to be treated for these in Canada and that doing so may lessen other patients' access to this medical care.

My Journey

1. *Did you know* that not all countries have the same patient and consumer protection laws we do here in Canada?
Before you decide whether to go, learn about malpractice laws in the countries you are considering visiting and make certain that your rights as a patient will be protected.

2. *Did you know* that some countries offer medical procedures to international patients that you cannot get here in Canada?
Before you decide whether to go, find out if the procedure you are considering has been shown to be safe and effective by a reputable agency.

3. *Did you know* that health care is a scarce resource around the world?
Before you decide whether to go, inform yourself about the most pressing health care challenges in the countries you are thinking about visiting and consider whether medical tourism may help or worsen them.

Be informed and share your thoughts, ideas, and experiences with others.
For more information go to: www.sfu.ca/medicaltourism/guide

Figure 2. Draft information sheet.

Data collection

Twenty-four semi-structured interviews were completed between October 2012 and December 2012. We conducted as many interviews as possible during this two-month period, after which we planned to cease data collection. The interviews were all conducted over the phone and recorded for future transcription. The shortest interview lasted just over 20 minutes, with the longest lasting nearly an hour and a half. The majority of interviews lasted approximately 45 minutes. The range in interview times is due to the semi-structured approach which provided some questions to guide the discussion and ensure meaningful feedback, but allowed for freedom for the participants to speak openly, which may broaden the understanding of the topic at hand [30].

All interviews were conducted by the first author using a semi-structured interview guide. The guide contained questions divided into four parts. The first part addressed basic background information of the participant including demographics, general health, and past travel experience. The second part asked questions relating to the participants' general decision-making behaviour. The third section asked questions related to participant's experiences as a medical tourist, particularly in terms of how they made the decisions relating to their trip. Both the second and third section probed the participant about the types of information they were considering during their decision-making process as a medical tourist. The final section of the interview guide contained questions asking for feedback on a draft version of the information sheet (see Figure 2), particularly the usefulness of such a document in the participants' context and recommendations for improving the document.

Data analysis

Interviews were transcribed verbatim. The first author reviewed all of the transcripts and extracted basic demographic information, basic procedure and trip information, as well as general feedback (positive or negative) on the information tool for each participant into a spreadsheet. She then assigned 6 transcripts for independent review by each team member. Three of these transcripts were the same across all members while three were unique. This strategy was used to minimize redundancy in transcript review and ensure that all members would encounter the range of opinions shared by participants. Following transcript review, we met to discuss emerging themes in participant feedback on the tool. Given the practical or applied nature of this formative study, the themes identified were primarily a priori or deductive in nature in that they focused on tangible improvements to specific aspects of the informational tool. After these themes were identified, the first author hand reviewed the transcripts and extracted segments that pertained to each into a Word document organized under six headings: positive feedback, negative feedback, specific recommendations, impacts of the tool on decision-making, potential places to access such an information tool, and type/tone of information conveyed in the tool. Verbal and written consent was obtained from interview participants for the use of anonymized quotes in published materials.

Results

Overall, the 24 interview participants ranged in age from 24 to 65. Nine participants identified themselves as male and the remainder as female. Fifteen of the participants had traveled out of Canada for the Chronic Cerebrospinal Venous Insufficiency (CCSVI) procedure, also called renal or jugular angioplasty by participants. The remaining nine participants accessed a lumbar disc replacement surgery, a dental surgery, a gastric band surgery, a tattoo removal through surgical incision, a vertical sleeve gastrectomy, an eye-lid surgery, a meniscus surgery, and a cholecystectomy. Five of the 24 participants indicated that they had existing family connections in the destination where they underwent the medical procedures, with four of these five participants having emigrated to Canada from that country. When examining destination choice amongst CCSVI patients, seven out of fifteen participants had traveled to the United States, with the remainder traveling to Bulgaria, Poland, Mexico, India, Germany, and Egypt. Non-CCSVI patients traveled to Egypt, Germany, Spain, Philippines, Venezuela, Thailand, Hong Kong, Romania, Mexico and India for surgery. Overall, five participants had participated in multiple medical tourism trips, with four out of five repeating the CCSVI procedure in a different location, and one travelling for three separate medical procedures in three different locations.

The 24 participants all read the information sheet prior to commencing the interview to allow for discussion on this document during the interview. The feedback provided by these participants is described in the remainder of this section. This feedback indicates the perceptions of these persons on the information sheet itself, with a particular focus on the format, usefulness and potential influence of the document on medical tourists' decision-making.

Feedback on format

Overall, almost all participants were generally positive when discussing the format of the tool. Regardless of the procedure sought or the country visited, participants had similar feedback on the format of the information tool, which was overwhelmingly positive. Positive feedback relating to format indicates that participants appreciated the concision of the document, including its short sentences and "*conversational*" language. Participants indicated that the document is "*easy to read*", relating this ease to both layout and language. Feedback on the format specifically indicates that the language is "*very clear*" and "*basic*". This was particularly important because many participants commented that the language and level of detail contained in the document must be easily accessible to persons with "*standard education*" and "*standard English speaking*" abilities. Furthermore, participants indicated that the short one-page format is favourable as the limited text is "*not boring*". The question-style format of the document was well-received by some participants who found this style encouraged persons to reflect on the information and take this information to do further research on their own, while others preferred that it be populated entirely with standard sentences.

While there were very few negative comments about the format of the document in terms of layout, participants provided formatting suggestions to improve the appeal of the document. Some participants recommended more of a checklist format over the question-style format. This format was recommended to avoid the document being "*wordy and plain*". Several participants also indicated that some of the repetition in wording was not appealing and made the document seem repetitive and condescending. The most common recommendation for improving the format of the document was to include pictures, images or more colours to increase the visual appeal.

In terms of language, while all participants agreed that the language was easy to understand, some participants pointed out areas where the information provided did not lend itself to specific actions. These participants demonstrated frustration with being unsure what the information in the tool was trying to convey. One participant indicated that education level would really impact whether a "*person's going to give it a second thought or not*" and suggested that while the language may be clear, the intended information may not be fully understood by all persons. Furthermore, some participants suggested that readers may have made up their mind before accessing the tool and may not have an open mind when reading the document, indicating challenges in conveying intended information through the format of an information tool.

Another suggestion by participants for maintaining the concision of the document while providing useful information was the inclusion of small anecdotes or stories as part of the format would be useful. Participants indicated that describing personal experiences through anecdotes helps empathize with the reader by creating a "*personal touch*". They felt that a format with anecdotes illustrating both positive and negative experiences would prompt persons to do further research while creating a neutral tone.

Feedback on usefulness

Positive feedback on the usefulness of the information in the tool was fairly congruent, with similar feedback coming from participants who had travelled to different countries for distinct procedures. In general, participants provided positive feedback on the usefulness of the document in prompting further research by those considering medical tourism. Many participants stressed the importance of medical tourists doing "*as much research as you can*" after reading the document and felt that this document is successful in "*giv[ing] you things to think about*" while engaging in this research. Some participants indicated specific examples of points of information from the document that they had not previously considered that encouraged serious contemplation when reading the tool. Participants that indicated the tool would be useful for promoting further contemplation and research of medical tourism generally felt that the information was relevant to anyone travelling out of country for medical care, regardless of procedure.

Only a small minority of participants indicated that they did not think the informational tool would be useful for Canadians considering medical tourism. There were two main bases for this perspective: 1) even if the document does prompt a reader to do further research, it does not provide enough guidance or insight into how to do this research, or 2) the information presented in the document is not relevant to their experience. Regarding the first basis, some participants indicated frustration after reading this document and wanting to know more about the issues being raised but not being sure of where to turn to access it. While the tool does provide a link to our own research website at the bottom of the page, some participants would have liked to have seen links to sources of information for each point.

Furthermore, some participants, particularly CCSVI patients, were unsure what to do if they wanted to go abroad for surgery that they were denied access to domestically in that they felt this circumstance raised unique issues. Overall, participants who were unsure as to how to do further research after reading the document felt that the tool was "*unrealistic*" as it lacked the sufficient amount of information necessary to make informed decisions regarding the points raised in the tool.

Participants who did not find the tool particularly relevant indicated that while they feel it is important to do a great deal of research before travelling as a medical tourist, often the motivations for travelling involve some level of desperation which really impacts upon the decision-making process. In other words, it was thought that this search for hope or push abroad out of desperation impacts the usefulness of the tool in that this reality far outweighs the importance of any of the prompts shared in the document. This desperation for medical treatment makes further research about the information contained in the document a "*bit of a tall order*". Many participants said that at the time of deciding whether or not to travel out of Canada for medical care, they were in a lot of pain and this would have impacted their abilities to spend time searching for additional information to guide them in their decision-making both in terms of energy levels and because it would be "*something else to worry about*".

Some participants did not agree with all of the potential negative impacts of medical tourism suggested in the tool. They were most critical of suggestions that local citizens may not benefit from medical tourism hospitals and clinics. These participants often cited witnessing locals being treated at the facilities they visited or the undeniable economic gain to the community as providing confidence that medical tourism does not negatively impact destination communities. One participant said that he or she was "*keeping the faith that us being there would provide more money*" into the local economy, indicating that she assumed an overall positive benefit of medical tourism on the local community. Likewise, some participants disagreed with points in the document related to potential negative impacts on Canada, indicating that leaving Canada for care would provide overall benefits such as "*reduce[d] waiting times*" and "*decrease[d] costs*".

Potential influence of the information tool on decision-making

When participants were asked whether or not the information tool would have impacted their decision-making, many indicated that they probably would have still travelled out of country for their medical procedure but that it would have prompted them into conducting more research before leaving Canada. They indicated that the document might specifically impact upon intended users' expectations by encouraging further research, the outcome of which would provide more realistic expectations when considering or planning a medical tourism trip. On the other hand, some participants clearly stated that they would not have considered the information in this tool, either because they perceived that the document was implicitly and explicitly biased against medical tourism that would have turned them off from reading the tool, or because they had already considered all of the information contained in this document.

When asked directly about ethical prompts included in the tool, most participants said they were not apparent to them in their review of the document. One participant said that it did not raise any ethical concerns to him or her because the tool "*is just stating a fact*" and "*not [presenting medical tourism] as either pro or con*",indicating that the perceived neutrality of the document impacted her or her perception on whether the information presents ethical concerns. The few participants that felt that the information in the tool raised ethical concerns did not see this tool as being explicit enough when raising these concerns, with one stating that "*the document doesn't really bring out the ethical concern of monopolizing [health] resources*" and another saying "*I don't see how this document raises those type of concerns; however, they're [concerns about patients in destination countries accessing medical] very valid concerns... that could be an ethical issue. But, if you bring it back to the document, I don't see how that's really linked to the type of questioning you ask in the document*".

DISCUSSION

The feedback provided by Canadians that had previously travelled for medical tourism demonstrate varied opinions regarding the format, usefulness, and potential impact of an information tool for Canadians considering engaging in medical tourism. Although there was not complete agreement amongst all participants in terms of their opinion of the tool, the findings shared above demonstrate recurring themes throughout the feedback. While feedback on tool format was overall quite positive, feedback on usefulness and impact of the tool was divided between a few different opinions that ranged from strongly positive to strongly negative. The positive feedback indicated that the concision of the tool and language accessibility were generally appealing to participants. Furthermore, the overall positive feedback on usefulness of the tool indicated that as a source of information that prompts further research and contemplation about medical tourism prior to coming to a final decision, this tool will be of interest to a wide range of persons considering medical tourism as it is not guiding any particular action. This positive feedback leaves us encouraged a broadly disseminated tool would be appropriate for the intended target audience and could result in increased awareness of ethical concerns of medical tourism.

Negative feedback about the tool identified potential barriers to increasing awareness of ethical concerns for medical tourism. First of all, strong negative feedback on the document focused on the perceived negative or biased tone of the document and indicated that this might dissuade people from considering the information presented in the document. However, we expect that participants may have been more likely to view the tone as negative given that they had already travelled as medical tourists and may be more likely to reject information indicating potential negative impacts of medical tourism or to feel defensive around the choice they made [12]. For example, many participants described in detail their experiences accessing high quality medical care as a medical tourist and this may have biased their view on the potential safety concerns related to this practice. Similarly, participants that disagreed with potential negative impacts on access to medical care for destination populations might have been impacted by their experiences with a particular community, which might not be the case for persons that have not already travelled as a medical tourist.

Negative feedback on the usefulness of the document did highlight a lack of 'know-how' for what to do next with the information provided in the tool. Even when participants acknowledged that this tool increased their awareness of potential impacts of medical tourism, many were unsure how this might translate to action. Participants experiencing chronic pain said that, given their desperation to access care, their focus on improving their individual quality of life would likely have prevented them from considering potential societal impacts had they read this document before travelling for care. These results are similar to previous studies that have found Canadian medical tourists are less inclined to consider potential societal impacts resulting from medical tourism when asked about ethical concerns, and more likely to discuss concerns about the Canadian health care system contributing to their decision to participate in medical tourism [14]. In fact, medical tourists are most likely to travel for medical care in response to their experiences of feeling 'abandoned' by their own health care system [31]. As the majority of participants in this study were diagnosed with multiple sclerosis and travelled outside of Canada to access an unavailable procedure, many participants' likely experienced high levels of frustration trying to access care in the Canadian health care system before deciding to travel outside of Canada for the CCSVI procedure. The interviews demonstrated that the suggestion to contemplate potential ethical and health equity impacts of medical tourism on the Canadian health care system or health care systems globally may further frustrate patients that feel abandoned by the health care system and are desperate to access care. These findings agree with previous research which describes medical tourists as escaping the constraints of a national health care system by expanding their health care system to a global scale [31].

Participants mentioned several times that their desperation resulted in quick decision-making around medical tourism, which may limit informed consent, an existing problem in medical tourism [2]. This tool intends to encourage further consideration of information regarding potential impacts of medical tourism but, as demonstrated in the results, this may be frustrating for participants who feel as though they are in a desperate situation, face time constraints, and/or are unsure where to begin this research. However, we believe this frustration, if voiced, could potentially have a positive impact by promoting a need for greater research and regulatory structures within the medical tourism industry that facilitate more ethical decision-making and informed consent by medical tourists. Research on policy development indicates that shaping the public debate may have a large impact on resulting policy outcomes [32] and an information tool such as this one can definitely play a role in shaping the public debate around medical tourism in Canada and increasing advocacy surrounding ethical medical tourism and health service provision and utilization.

Following analysis of feedback from interview participants, the research team utilized this feedback to further revise the information provided in the document. Furthermore, we took into consideration feedback on document format from the interview participants as well as consultation with existing one-page health communication documents provided to patients in travel health clinics to guide the design of a document for dissemination. This final document can be seen in Figure 2. Dissemination of this document will take place through distributing copies to travel health clinics, primary health care providers, health authorities, and other groups interested in providing this document to persons using their services.

Wider relevance

The results and discussion of feedback from interview participants provide insight into the challenges and useful considerations that researchers, policy makers, and health care providers may wish to contemplate when considering the communication of ethics in health care utilization and provision. Global health researchers have identified a need for the application of ethical frameworks and tools to public health to encourage more fair and just allocation of health resources globally [33,34]. Consumer guidelines utilize a format that has been effective at increasing public discussion about some of the existing ethical issues in practices such as tourism. This public discussion may serve to introduce and reinforce more ethical practices through increased awareness of one's social responsibilities [35]. This research study demonstrates that guidelines or information sheets tailored to health care users may provide an effective means of engaging persons accessing health care into conversations and further contemplation about one's social responsibilities when accessing health care. As such, this research may be of interest to persons interested in promoting more just and fair allocation of health resources.

Limitations

This study is limited in its ability to develop an understanding of persons' actions following the consideration of information provided in this document. As the methods utilized for this analysis involved one-time interviews with persons that had already travelled for medical tourism, these persons did not speak on any future actions that they might take. While this study explored contexts in which this information may not be considered by someone reading the document, for those that did consider this information useful to the decision-making process, this study is limited in its ability to provide insight into the actual use of this information in this decision-making process. As such, the ability for this study to understand the effectiveness of the information sheet is limited.

Future directions

While the findings of this formative study suggest that an information sheet focused on ethical concerns with medical tourism can play a useful role in raising awareness about these concerns and advocating for more ethical practice in the industry regardless of procedure, more research is needed to further explore this potential given the limitations of this study. For example, feedback from persons that have not already travelled as medical tourists could provide additional insight on the potential impact of this information tool. Additionally, there was limited feedback provided by participants on their understanding or potential engagement with the information provided in the tool. While participants easily engaged in conversation about potential impacts of medical tourism in previous questions, when asked about their thoughts on whether the information in this tool presented them with any ethical concerns, the use of the word "ethics" seemed to limit participants' engagement. This may be due to participants feeling uncomfortable discussing the ethics of an activity in which they have already engaged. This more limited discussion of ethics resulted in some lack of clarity regarding comprehension of the information presented in the tool. This suggests that future research with persons contemplating medical tourism should focus on greater in depth-discussion about the impacts of such a tool in terms of engagement with the information presented within. Furthermore, an additional resource such as a website could provide greater details, including anecdotes and alternative methods of describing the ethical concerns of medical tourism. The link to the website could be provided on the information sheet, and further research could explore feedback on the usefulness of such a website.

CONCLUSIONS

This paper has explored the use of an information tool as a means of increasing individuals' awareness of ethical concerns in medical tourism to enable Canadians to make more informed decisions about private health care utilization abroad. With existing sources of information for medical tourists commonly demonstrating a lack of neutral guidance [12,16], this information tool responds to a pressing, practical knowledge gap. According to feedback from interviews with former Canadian medical tourists, the tool we have developed has the potential to raise awareness of ethical concerns during the decision-making process. However, it is unknown at this point whether this awareness will lead to shifts in attitude and changes in behaviour at both the individual and societal level to contribute to social responsibility in health care provision and utilization, and ultimately improved global health equity [28]. Moving beyond the formative research presented herein, implementation of the tool and evaluation of its uptake and utilization will shed light on whether or not such awareness will lead to these types of shifts and changes.

COMPETING INTERESTS

The authors declare that they have no competing interests.

AUTHORS' CONTRIBUTIONS

KA conducted 23 of the interviews while JS conducted 1 interview. KA reviewed all interview transcripts and the remaining authors reviewed 6 transcripts each. All authors contributed to the development of the interview guide, discussions of emerging findings from the interviews, and revisions to the information sheet. KA wrote the manuscript with input from all other authors. JS and VAC reviewed and approved the final manuscript. All authors read and approved the final manuscript.

ACKNOWLEDGEMENTS

The development of this paper was supported through a Catalyst: Ethics Grant from the Canadian Institute of Health Research (funding reference number 114797). VAC is also supported by a Scholar Award from the Michael Smith Foundation for Health Research.

REFERENCES

1. Johnston R, Crooks VA, Snyder J, Kingsbury P. What is known about the effects of medical tourism in destination and departure countries?: A scoping review. International Journal for Equity in Health.2010;8:24.

2. Penney K, Snyder J, Crooks VA, Johnston R. Risk communication and informed consent in the medical tourism industry: a thematic content analysis of Canadian broker websites. BMC Med Ethics.2011;8:17.

3. Snyder J, Crooks VA, Johnston R, Kingsbury P. Beyond sun, sand & stitches: assigning responsibility for the harms of medical tourism. Bioethics. 2012;8:233–242.

4. Lunt N, Carrera P. Systematic review of web sites for prospective medical tourists. Tourism Rev.2011;8:57–67. doi: 10.1108/16605371111127224. 5. Advisory Committee on Genetic Modification:*Annual Report.* London; 1999.

5. Crooks VA, Turner L, Cohen G, Bristeir J, Snyder J, Casey V, Whitmore R. Ethical and legal implications of the risks of medical tourism for patients: a qualitative study of Canadian health and safety representatives' perspectives. BMJ Open. 2013;8:2.

6. Connell J. Contemporary medical tourism: conceptualisation, culture and commodification. Tour Manage. 2013;8:1–13.

7. Hopkins L, Labonte R, Runnels V, Packer C. Medical tourism today: what is the state of existing knowledge? J Public Health Policy. 2010;8:185–198.

8. Pocock NS, Hong Phua K. Medical tourism and policy implications for health systems: a conceptual framework from a comparative study of Thailand, Singapore and Malaysia. Globalization and Health.2011;8:12.

9. Horowitz MD, Rosensweig JA, Hones CA. Medical tourism: globalization of the healthcare marketplace. Medsc Gen Med. 2007;8:33.

10. Widdows H. Localized past, globalized future: towards an effective bioethical framework using examples from population genetics and medical tourism. Bioethics. 2011;8:83–91.

11. Ormond M. Shifting subject of health-care: place 'medical tourism' in the context of Malaysian domestic health care reform. Asia Pac Viewp. 2011;8:247–259. doi:

12. Johnston R, Crooks VA, Snyder J. “I didn’t even know what I was looking for...”: a qualitative study of the decision-making processes of Canadian medical tourists. Asia Pac Viewp. 2012;8:23.

13. Turner L. ‘First world care at third world prices’: globalization, bioethics and medical tourism.BioSocieties. 2007;8:303–325.

14. Snyder J, Crooks VA, Johnston R. Perceptions of the ethics of medical tourism: comparing patient and academic perspectives. Ethics: Public Health; 2012. pp. 1–9.

15. Snyder J, Crooks VA. New ethical perspectives on medical tourism in the developing world. Dev World Bioeth. 2012;8:1.

16. Ormond M, Sothern M. You, too, can be an international medical traveler: reading medical travel guidebooks. Health Place. 2012;8:935–941.

17. Crooks VA, Bristeir J, Turner L, Snyder J, Casey V, Johnston R. Understanding the Health and Safety Risks for British Columbia’s Outbound Medical Tourists. Department of Geography: Simon Fraser University; 2011.

18. Terry N. Under-regulated healthcare phenomena in a flat world: medical tourism and outsourcing.New Engl Law Rev. 2007;8:415–463.

19. Walley J, Lawn JE, Tinker A, de Fransisco A, Chopra M, Rudan I, Bhutta ZA, Black RE. Primary health care: making alma-Ata a reality. Lancet. 2008;8:13–19.

20. Ostlin P, Schrecker T, Sadana R, Bonnefoy J, Hertzman C, Kelly MP, Kjellstrom T, Labonte R, Lundberg O, Muntaner C, Popay J, Sen G, Vaghri Z. Priorities for research on equity and health: towards an equity-focused health research agenda. PLOS Medicine. 2011;8:11. [

21. Sen Gupta A. Medical tourism in India: winners and losers. Indian J Med Ethics. 2008;8:3–5

22. Lunt N, Hardey M, Mannion R. Nip, tuck and click: medical tourism and the emergence of web-based health information. Open Med Informat J. 2010;8:1–11.

23. Chapman A. Health Care Reform: A Human Rights Approach. Washington, DC: Georgetown University press; 1994.

24. Marmot M, Allen J, Bell R, Goldblatt R. Building the global movement for health equity: from Santiago to Rio and beyond. Lancet. 2012;8:181–188.

25. Von Schirnding Y. Health and sustainable development: can we rise to the challenge? Lancet.2002;8:581.

26. Graces J, Rodenas F, Sanjose V. Towards a new welfare state: the social sustainability principle and health care strategies. Health Policy. 2003;8:201–215.

27. Parvanta C, Nelson DE, Parvanta S, Harner RN. Essentials of Public Health Communication. Jones & Bartlett Learning: Mississauga; 2011.

28. Bernhardt JM. Communication at the core of effective public health. Am J Public Health.2004;8:2051–2053.

29. Lefebvre RC, Flora JA. Social marketing and public health intervention. Health Educ Behav.1988;8:299–315.

30. Denzin NK, Lincoln YS. The SAGE Handbook of Qualitative Research. Thousand Oaks, California: SAGE Publications; 2011.

31. Perfetto R, Dholakia N. Exploring the cultural contradictions of medical tourism. Consumption Markets & Culture. 2010;8:399–417.

32. Waddell C, Lavis JN, Abelson J, Lomas J, Sheperd CA, Bird-Gayson T, Giacomini M, Offord DR. Research use in children's mental health policy in Canada: maintaining vigilance amid ambiguity. Soc Sci Med. 2005;8:1649–1657.

33. Farmer P. Pathologies of Power: Health, Human Rights, and the New War on the Poor. California: University of California Press; 2005.

34. Maclean SJ, Brown SA. The social determinants of global health: confronting inequities. In Health for Some: the political economy of global health governance. Edited by Maclean SJ, Brown SA, Fourie P.Palgrave Macmillan: Basinstoke; 2009.

35. Clarke N, Barnett C, Cloke P, Malpass A. Globalising the consumer: doing politics in an ethical register. Polit Geogr. 2007;8:231–249.

10

CHAPTER

Fly-By Medical Care: Conceptualizing the Global and Local Social Responsibilities of Medical Tourists and Physician Voluntourists

Jeremy Snyder[1*], Shafik Dharamsi[2] and Valorie A Crooks[3]

[1]*Faculty of Health Sciences, Simon Fraser University, Blusson Hall 11300, 8888 University Drive Burnaby BC, Canada.*

[2]*Department of Family Practice, University of British Columbia, David Strangway Building, 3rd Floor 5950 University Boulevard, Vancouver, BC. Canada.*

[3]*Department of Geography, RCB 6141, Simon Fraser University 8888 University Drive Burnaby, B C,Canada.*

ABSTRACT

Background

Medical tourism is a global health practice where patients travel abroad to receive health care. Voluntourism is a practice where physicians travel abroad to deliver health care. Both of these practices often entail travel from high income to low and middle income countries and both have been associated with possible negative impacts. In this paper, we explore the social responsibilities of medical tourists and voluntourists to identify commonalities and distinctions that can be used to develop a wider understanding of social responsibility in global health care practices.

Discussion

Social responsibility is a responsibility to promote the welfare of the communities to which one belongs or with which one interacts. Physicians stress their social responsibility to care for the welfare of their patients and their domestic communities. When physicians choose to travel to another county to provide medical care, this social responsibility is expanded to this new community. Patients too have a social responsibility to use their community's health resources efficiently and to promote the health of their community. When these patients choose to go abroad to receive medical care, this social responsibility applies to the new community as well. While voluntourists and medical tourists both see the scope of their social responsibilities expand by engaging in these global practices, the social responsibilities of physician voluntourists are much better defined than those of medical tourists. Guidelines for engaging in ethical voluntourism and training for voluntourists still need better development, but medical tourism as a practice should follow the lead of voluntourism by developing clearer norms for ethical medical tourism.

Summary

Much can be learned by examining the social responsibilities of medical tourists and voluntourists when they engage in global health practices. While each group needs better guidance for engaging in responsible forms of these practices, patients are at a disadvantage in understanding the effects of medical tourism and organizing responses to these impacts. Members of the medical professions and the medical tourism industry must take responsibility for providing better guidance for medical tourists.

BACKGROUND

The concept of social responsibility has been influential in guiding professionals' conduct, including in business [1,2], law [3, 4], and medicine [5, 6]. We understand social responsibility to entail the claim that an individual or group of individuals has a moral responsibility to promote the welfare of the communities to which they belong or with which they interact [6, 7]. For businesses, for example, corporate social responsibility is the claim that corporations have a responsibility to promote the welfare of the communities with which they do business, including providing a living wage to their employees, operating in an environmentally sustainable manner, and ensuring that some of their profits benefit community stakeholders [8]. Similarly, lawyers have not only a fiduciary responsibility to their clients, but also, as members of a profession, are obligated to engage in *pro bono* legal work that aids community members who are unable to pay for their services [9]. And for members of the medical profession, there have long been calls for physicians to look beyond the good of their own patients and act also to promote health within their communities [10].

Typically, calls for social responsibility focus on an obligation to promote domestic welfare. However, as individuals participate in more globally-oriented practices, the scope of the targets of their social responsibility expands. This phenomenon is evident in the corporate social responsibility literature focusing on multinational corporations [8,11]. For multinational corporations, their social responsibility is not discharged simply by benefiting the communities in which their corporate headquarters are located. Rather, they must also ensure that stakeholders in all of the communities in which they operate benefit from their operations and that these benefits are sustainable over the long term. In practical terms, this might mean that multinationals that outsource manufacturing from their home countries must ensure that they pay a living wage to foreign-based workers, refrain from polluting foreign communities, and spread some of their profits both at home and abroad [12, 13].

In this article, we explore the social responsibility of the participants in two global health care practices: voluntourism (travel abroad by physicians to deliver medical care) and medical tourism (travel abroad by patients to receive medical care). The terms 'voluntourism' and 'medical tourism' can both be seen as pejorative and normatively loaded given the connotation that each involves a frivolous, touristic element. For this reason, for example, some members of the medical tourism industry prefer labels such as 'medical travel' or 'global health care'. We use the terms 'voluntourism' and 'medical tourism' here because they are widely recognized and used in the academic literature. We do not intend to imply by the use of these labels that either practice is inherently morally problematic or any other related negative value judgments. While both of these groups share many attributes [14], we understand them to be distinct phenomena practised by different groups. We specifically aim to explore the nature of the social responsibilities of these two groups and to draw together parallels and distinctions that can be used to assist with articulating wider trends regarding social responsibility in global health care practices. In doing this we extend the traditional professional-centric focus of the social responsibility literature to consider the types of responsibilities inherent in the practice of medical tourism for international patients. While medical tourists travel from both developed and developing countries and represent a diverse range of income levels [15], we focus on the social responsibilities of relatively wealthy patients from high income countries in order to draw a parallel between the relative privilege of these patients and that of physician voluntourists traveling from high income countries. As we argue below, the better defined social responsibility of physicians engaging in voluntourism holds lessons for the rapidly developing practice of medical tourism. To accomplish our aim, we first provide an overview of the global practices of voluntourism and medical tourism, and then move to articulate the social responsibilities of voluntourists and medical tourists separately, focusing on the basis for their social responsibility and the targets of this responsibility. We then offer a discussion that compares these two groups, looking for overlaps and distinctive elements in their social responsibilities.

Global Health Care Mobilities: Introducing Voluntourism and Medical Tourism

Recent years have witnessed the emergence of new forms of global health care mobilities, and increased popularity of existing forms due to processes such as the development of a globalized economy, establishment of international and bi-lateral trade agreements, and expansion of the international travel industry [16, 17, 18]. Patient mobilities (the movement of patients across international borders for service use) and provider mobilities (the flow of health care providers across international borders for service delivery) are two important forms of international health care mobility. These mobilities take many forms and involve flows between an almost endless number of home countries and destination nations. Provider mobilities can include permanent health worker migration and short-term relocation to enhance skills through training abroad [16, 19], while patient mobilities can include accessing arranged cross-border care through referral and obtaining emergency care while abroad [17,20]. In the remainder of this article we focus on two specific forms of patient and provider mobility. Physician voluntourism and patient medical tourism are international health care mobilities that are both characterized by temporally-limited time abroad and engagement in a minimum of two health care systems, either as a user or provider.

Global health disparities and inequitable access to health care in developing countries is an ongoing concern for many physicians. For instance, sub-Saharan Africa has close to 25% of the global disease burden but has only 3% of the global healthcare workforce [21]. Globalized processes have enabled physicians from around the world, and particularly from high income countries, to participate in humanitarian "medical missions" to developing countries to administer medical care as physician volunteers [22]. Physician participants in these missions see themselves as part of a long-standing humanitarian tradition in medicine of bringing desperately needed medical care to vulnerable communities in developing countries. The popularity of medical volunteering is on the rise, with over 500 medical mission organizations in the United States alone that help to organize over 6000 short-term missions to foreign countries [23]. Medical students are also enrolling in increasing numbers to participate as volunteers in global health initiatives during their training. Current figures suggest that close to 30% of graduating North American medical students have taken part in a global health project [23]. Yet, there is also growing concern around the lack of ethical guidelines supporting medical missions and volunteerism that has resulted in the labelling of these terms as "physician voluntourism", used pejoratively to describe volunteering as initiatives that can do more harm than good [24, 25, 26, 27]. Nevertheless, those who continue to participate in this practice see it as a social responsibility and a form of global citizenship [28].

Medical tourism, on the other hand, takes place when patients leave the country in which they live to pursue non-emergency medical interventions abroad [20, 29]. The care accessed abroad is not part of an established cross-border care arrangement (e.g., does not involve physician referral), and is typically paid for out-of-pocket [20]. Medical tourism is thought to be a popular option for patients on wait-lists for care in their home systems, who have no health insurance or are underinsured, or who are looking to access experimental or illegal treatments [18,30, 31]. A number of developing countries, including India and Thailand, have become leaders in this international industry [32]. Unfortunately no reliable estimates exist regarding the number of people travelling abroad each year as medical tourists [29]. Despite this, estimates regularly project growth in the industry in the years to come [20]. With the growth of the industry have also come concerns regarding the impacts it is having on destinations, particularly within developing nations. An oft-repeated worry is that it will exacerbate health inequities in both the destination and home country for medical tourists [20, 29]. In the destination country, if medical tourists drive demand for expensive services, they may price out poorer citizens, or at least create a second tier of medical care in those countries [33, 34]. Medical tourism may shift services from preventive public health measures to less effective, and more expensive, clinical interventions [35]. The development of private clinics serving foreigners may also encourage the movement of trained physicians from the public to private sphere [33, 35]. On the other hand, proponents of medical tourism note its potential to cross-subsidize health care in the public sphere [36], though some of these agreements have been violated in practice [37].

FINDINGS

Physicians' Social Responsibility in Voluntourism

Physicians have long embraced a fiduciary duty to manage and protect the health of patients, over and above their own self-interest. This fiduciary relationship plays a foundational role in medicine, and is founded on principles such as fidelity, integrity, compassion, courage, altruism, and justice [38]. The concept of social responsibility is informed by these principles, and is one that enables physicians to develop a public trust, and a professional identity around what it means to be a Doctor in society. There is a sense that modern day medicine is failing to recognize its societal role [39] and failing to educate physicians to meet the health care needs of a diverse society [20, 40]. This situation is problematic as a physician's social responsibility to protect public interest is not an option, but a fiduciary duty that is entrusted to each and every physician, individually and collectively. It is based on the understanding that illness affects an individual's capacity to function as a productive and contributing citizen, member of a family unit, and part of the socio-economic system. Health and health care, therefore, are regarded by many countries as concerns of society as a whole and not simply those who are ill. In the remainder of this section we explore the various dimensions of physicians' social responsibility and consider how they relate to their involvement in the global health care practice of voluntourism.

Physicians' Social Responsibility

One manifestation of physician's social responsibility is the obligation to respond to inequities in health and how health services are organized in their domestic communities. It requires physicians to be mindful of responsibilities beyond individualism, profit, and private interests. The first Code of Ethics, issued by the American Medical Association in 1847, defines the duties of physicians to their patients, to each other, and to the general public:

As good citizens, it is the duty of physicians to be ever vigilant for the welfare of the community, and to bear their part in sustaining its institutions and burdens: they should also be ever ready to give counsel to the public in relation to matters especially appertaining to their profession[41].

Physicians are called upon to safeguard health systems so that services are effective, efficient, equitable, and sustainable [42]. Social responsibility is not simply a matter of charity, but a moral commitment to the patient that has been developed over centuries within societies that have advanced the conception of medicine as a profession. Professional status cannot be claimed without public sanction [43, 44, 45]. For this reason, physicians are required to maintain very high levels of expertise and skillfulness, as well as virtuousness and trustworthiness.

The provision of health care as a social security measure within an organized social system dates back to early Egyptian and Greek civilizations where physicians were hired by the state to treat its citizens without charge [46]. In 1601, Britain passed the Elizabethan Poor Law, allowing for a general taxation system to ensure medical care for the poor and infirm; and during the latter part of the Industrial Revolution several social, professional and religious associations or guilds also contributed a set sum of money voluntarily toward a form of protection that could provide assistance to its members who became incapacitated due to illness [47]. These early initiatives established a precedent regarding physicians' involvement in maintaining the social, or common, good beyond simply caring for their patients.

The notion of health care as a common good, rooted in social and religious ideas of charity, beneficence and compassion, is now recognized within the broader context of distributive justice, and the growing sensitivity to the equitable distribution of health care [48]. Hence, in medicine there is a growing reaffirmation that physicians have an obligation to the individual patient as well as an enduring responsibility to the broader society [49], particularly when dealing with issues around resource allocation, the social determinants of health, and related inequities. To this effect, the World Health

Organization suggests that physicians need to be mindful of medicine's social responsibility [50]. Hence, medical organizations are called to direct their education, research and service activities toward addressing the priority health concerns of each community, region and/or nation that they have a mandate to serve, and particularly the more vulnerable and marginalized segments of their populations [51].

Medical Voluntourism and Physician Social Responsibility

Though discussions of physicians' social responsibility tend to focus on their responsibility to domestic communities, many medical students and physicians choose also to participate in medical voluntourism abroad out of a sense of social responsibility [21, 28, 52, 53]. Through the act of voluntourism, these physicians invest personal time and resources toward reducing global health inequities. However, the growth of medical voluntourism is also outpacing the development of physicians' social responsibilities toward communities abroad and ethical guidelines to ensure that vulnerable communities are not subjected to more harm than good. Concerns about the lack of guidance for voluntourists derive in part from ethical tensions that emerge when research projects are conducted by researchers from high income countries in developing countries [24]. While host country members appreciate some aspects of these volunteers' work, responses to voluntourism are mixed [54].

A short-term clinical stint in a developing country can be seen as nothing more than a glorified form of tourism wrapped in a veneer of altruism, with no sustainable benefits for receiving communities [55, 56, 57]. Medical student voluntourists have also been criticized for using vulnerable people in developing countries to practice clinical skills, enhance résumés, and provide opportunities for travel to far-away and exotic places. Shah and Wu [58] provide a compelling example of the possible negative outcomes of medical student voluntourism through sharing a student's reflection:

After finishing my first year of medical school, I participated in a mission trip to Mexico. Before flying to Mexico, I was not given any cultural, medical, or other training, nor could I speak Spanish. Upon arriving, I was assigned to a clinic where there were hundreds of patients but only one physician. I remember vividly seeing a frail 11-year-old boy with polyuria, polydipsia and nocturia. My lack of medical training limited my differential. With only a scattered history and no other tests, I told him to limit caffeine intake and see if that helps. Thinking back, he could have had a urinary tract infection, any number of renal abnormalities, or worse, I sent him out without ruling out diabetic ketoacidosis. And while I was seeing patients by myself, other first year medical students were performing surgeries in the other clinics and later bragging about it.

The bragging by these students highlights the danger of voluntourism serving the needs of the voluntourist rather than the community abroad. Providing health care in international settings without carefully thinking about patient safety, sustainability, cultural appropriateness, quality of care, and consultation with local healthcare providers, among other similar issues, threatens to run counter to rather than discharge physicians' social responsibility abroad. Although participation in global health initiatives has great potential to offer medical trainees and physicians the opportunity to discharge their social responsibility [24], the risk of undesirable impacts from voluntourism can outweigh these benefits [59]. Vulnerable communities can easily become a means to the volunteers' ends instead of serving first the community's identified needs and empowerment interests.

Voluntourism is also often criticized for taking an exclusively charity-based approach to the provision of medical care, rather than enabling an equal and collaborative partnership with communities for developing capacity to address the root causes of systemic social inequity and disparity [60]. Charity based activities are based on the "good Samaritan" concept - providing resources, time, knowledge, and clinical service to vulnerable people. This approach is not only difficult to sustain, it can also create a dependency relationship through 'band-aid' solutions that do not address the root problem of health disparities. This line of criticism of voluntourism parallels critiques centred on the common establishment of temporally-limited selective primary health care initiatives in developing nations through aid programs, where a more community-centred intervention is thought to be the creation of long-term comprehensive primary health care plans [61, 62]. In relation to voluntourism, a sustainable and community-centred approach requires physicians to focus their efforts on understanding and working to change the structural or institutional factors that contribute to inequitable conditions.

The Association of American Medical Colleges' (AAMC's) offers four foundational ethical considerations prior to embarking on global health voluntourism: (1) ensuring high ethical and moral standards, (2) developing a social contract with the communities served, (3) subordinating self-interest to the interest of the communities served,

Table 1. The 4Rs of Ethically Sound Research

Ethical Principle	Strategy
Respect	Valuing cultures' and communities' diverse knowledges regarding health matters and developing knowledge that contributes to communities' and cultures' health and wellbeing
Relevance	Ensuring that research (or practice) is relevant to the culture and community
Reciprocity	Incorporating a two-way process of knowledge exchange and learning, where all parties benefit from these opportunities and the development of relationships
Responsibility	Fostering empowerment through allowing for active participation and rigorous engagement by all parties

and (4) ensure that core humanistic values (honesty and integrity, caring and compassion, altruism and empathy, respect for self and others) are at the forefront of all activities [23]. These ethical considerations point to a number of specific social responsibilities that physicians involved in voluntourism hold, such as ensuring that compassionate and respectful care is provided that meets the highest ethical and moral standards that the context allows for. What these guidelines lack are specific, concrete strategies for enacting ethical, socially responsible care. The 4Rs that were developed by Aboriginal leaders in Canada to guide researchers in working with their communities, which are summarized in Table 1, offer some suggestions for specific strategies [63].

Generally, socially responsible medical voluntourism is a collaborative process that considers the full participation of local communities, local healthcare workers, and local health authorities [54]. It complements principles of international solidarity and social capital within the context of civil society, where voluntourists act voluntarily and without seeking personal profit to share benefits.

Patients' Social Responsibility in Medical Tourism

While patients do not form a professional group, with their own institutions, leadership, and codes of ethics like physicians, there have been claims that individual patients do have social responsibilities to their domestic communities. Much of the literature on patient responsibility has focused on the degree to which patients are responsible for their own health [64]. This literature seeks to determine the balance between personal responsibility for health and the responsibility of communities for the health of their individual members. There is, however, some discussion of the responsibilities of patients to their domestic communities and to their health care systems [65, 66]. In the remainder of this section we articulate the hallmarks of patients' social responsibility and consider the specific types of responsibilities international patients hold when they engage in medical tourism.

Patients' Social Responsibility

Patients may have a sense of social responsibility due to having a sense of solidarity among the members of a community (e.g., other clinic users). Solidarity can represent a sense of togetherness and independence between individuals. Community members need not feel personally close or attached, but rather are part of a system that is valuable. These systems are made up of shared institutions, an example of which is a health care system. For individuals, this sense of solidarity implies not simply that the individual receives benefits from these institutions, but that she also contributes back in keeping with a value of reciprocity. In the context of solidarity around institutions that provide for the health of a community, "people should not be only passive recipients of services but should actively contribute to and try to avoid harming the system. This means that they should act responsibly when it comes to their health and that it is justified to expect this to a certain reasonable degree" [66]. Without reciprocity, shared institutions are unlikely to survive and the shared good will be lost. On this reading of personal responsibility, looking after one's own health and the efficient use of public health care resources can be understood as an expression of solidarity with community members. In addition to responsibilities for one's own health, the patient may also be said to have responsibilities to: (1) others, in the form of not harming others and meeting the health needs of those under one's guardianship; (2) the health care system, so that it may function fairly and efficiently and serve as many people as fairly as possible; and (3) the judicial authority, where patient responsibilities have been codified explicitly [65, 67].

Under public health care schemes, patients have a responsibility to look after their own health for their own sake, but also as a social responsibility to the other contributors to the health system and to the health system itself. For example, the Romanow Report in Canada includes a proposed health covenant that lists a series of responsibilities for Canadians, including to "observe good health practices, and to promote and support the well-being of their families and communities" and "to use the system prudently, and to support the system through their actions and tax dollars" [68] (*p.50*). Similarly, the National Health Service (NHS) in Scotland distributed a pamphlet called *The NHS and You*[69] that details both the responsibilities of the NHS to its patients and the responsibilities of patients to the NHS. These responsibilities are clearly directed toward the wider community and the system itself, as they are ways that the patient can help "yourself, other patients, and NHS staff" [69] (*p. 15*). These responsibilities include: treating NHS staff considerately, keeping appointments and informing staff if an appointment must be cancelled, keeping contact information up to date, following medical advice, using emergency services appropriately, finishing any course of medications, and helping to stop the spread of infection. The pamphlet also discusses other ways to help promote health, including by donating blood, organs, and tissues and by volunteering with the NHS. These responsibilities are intended to allow the public system to operate more efficiently and better serve the whole community.

While the Canadian and Scottish examples above are non-binding, a Medicaid member agreement in the US state of West Virginia is binding on its members. Some of the responsibilities listed in this document are responsibilities to look after the patient's own health, though these responsibilities too can be construed as a social responsibility to use public resources efficiently. Other listed responsibilities are more clearly injunctions against inefficient use of public resources. These responsibilities include requirements to show up on time for appointments ("I will show up on time when I have my appointments" and "I will bring my children to their appointments on time"), the responsibility to facilitate contact with the Medicaid system ("I will let my medical home know when there has been a change in my address or phone number for myself or my children"), and the responsibility not to misuse emergency services ("I will use the hospital emergency room only for emergencies") [70]. Similarly, the state Medicare program in Kentucky includes the interlinked goals to "1) Stretch resources to most appropriately meet the needs of members; and 2) Encourage Medicaid members to be personally responsible for their own health care" [71]*(p.3)*. As with West Virginia, the Kentucky plan targets additional 'get healthy' benefits to persons who document participation in identified healthy practices.

The guidelines shared above have been rightfully criticized as potentially shifting burdens onto the most vulnerable members of society as Medicaid users in the US fall into the lowest income brackets [72, 73]. These concerns can be addressed by noting that the patient's social responsibility is coupled with society's responsibility to provide for community health and limited by the patient's capacity for choice. That is, we can describe the responsibilities of society to patients, particularly for the social determinants of health, while at the same time acknowledging the role of personal conduct not only in personal health, but also in the functioning of one's health care system. This mutual responsibility for health admits of degrees just as an individual's ability to control her health varies depending on contextual factors, including her position in her social hierarchy [65].

Medical Tourism and Patient Social Responsibility

If medical tourists have a social responsibility to look to the efficient functioning of their own domestic health systems, then participation in medical tourism will extend this responsibility to the health systems of the destination countries to which they travel and develop new connections. Medical tourism for procedures that will serve to undermine health equity and the sustainability of the health system in destination countries is therefore a potential violation of the patient's social responsibility. Crucially, however, many of the worries about the negative impacts of medical tourism on destination countries are matters of conjecture rather than well-established fact [29]. Moreover, while many instances of medical tourism may exacerbate health inequities, it is not clear that all forms of medical tourism are fated to do so. Medical tourists who wish to engage in forms of medical tourism that do not cause these negative effects for destination countries, then, will be faced with severe difficulty in assessing the effects of their travel.

Medical tourism has also been associated with negative effects for the patient's home country in terms of lessening equitable access to care. As medical tourism allows relatively wealthy patients to opt out of treatment in their home health care systems, it may undermine political pressure for change as privileged patients are able to have their health care needs met abroad [74]. If so, less privileged patients who are less mobile will be left in a lower tier of care at home. For publically funded health care systems, the practice of paying out of pocket for necessary medical services can also help to encourage the privatization of health services at home, which may also undermine health equity [75, 76]. As with the negative effects of medical tourism on destination countries, however, these concerns are mostly matters of conjecture. While the patient may have a social responsibility not to travel abroad for care if doing so will undermine efficiency and equity in her home system, she may not have the information necessary to judge whether becoming a medical tourist will encourage these effects.

Medical tourism can serve as a means for patients to secure care more cheaply and quickly than if they remain within their local health care systems. However, travelling abroad for care creates a series of risks for the patient and long-term costs for the patient's home health system [20]. Travel itself creates risks by hastening the pace of care and surgeries and by increasing the risk of deep vein thrombosis or other complications from long plane flights [77]. Travel for care abroad can result in negative health consequences if an experimental treatment results in complications for the patient or other side effects [78]. While care abroad, even in low and middle income countries, is often of very high quality, poor oversight of facilities in some countries can result in sub-standard care and therefore complications and the need for follow-up care for the patient [79]. Patients receiving care abroad may also bring infections back home with them, including the NDM1 'superbug' that has been linked to medical tourists [80]. Finally, many forms of treatment require extensive follow-up care even if the principle intervention is successful or completed without complications. If arrangements for follow-up care in the patient's home country have not been made, then recovery can be delayed, resulting in complications [79]. Similarly, difficulties in transferring medical records between home and destination countries can complicate follow-up care [81].

As a result of these risks for medical tourists, they may incur more extensive expenses for follow-up care than persons remaining within their home countries. Insofar as these patients have a social responsibility to look to the efficient functioning of their home health care systems, engaging in medical tourism can potentially constitute a failure to discharge this social responsibility. Such an efficiency-based responsibility has been codified in Germany, for example. There, patients are asked to respect "the clinical and cost effectiveness of services, which are only to be used insofar as necessary" [67]. These responsibilities will exist for both members of public systems like those in the UK, Canada, and Germany, public portions of highly privatized systems like Medicaid in the US, and even insurance holders in privatized systems who have a duty of solidarity to their fellow insurance pool members.

As with other patient social responsibilities, the responsibility to use health resources efficiently should not undermine fair access to care, should not fall disproportionately on disadvantaged populations, and must admit of degrees in reflection of the extent of individual choice over health care decisions [67]. As some patients engage in medical tourism for necessary care that they would not otherwise be able to afford or access, their decision to go abroad for care may not be a matter of choice. Moreover, patients may not be aware of the dangers associated with medical tourism or the requirements for follow-up care for their specific procedures [75, 82]. More generally, discerning the effects of engaging in medical tourism is difficult even for highly informed patients given a lack of data on the effects of medical tourism [29] and most patients would likely not have access to this information even if it were available. Therefore, it is inappropriate to hold medical tourists socially responsible for specific negative effects of medical tourism under these conditions. This is because discharging one's social responsibility by using health resources efficiently and mitigating third party harms requires knowledge of the effects of personal and health care choices [66]. Therefore, a first step toward a call to greater social responsibility among medical tourists is not to blame them for the effects of engaging in this global health care practice but rather to educate them on the effects of medical tourism.

In terms of assigning social responsibility for medical tourism, it is useful to differentiate between travel for medically-necessary and elective treatments. While many medical tourists travel abroad for much needed hip replacements, cardiac surgeries, or eye surgeries, other treatments such as elective cosmetic surgery would not be considered medically necessary. While we can grant that there will be considerable grey area between the categories of medically necessary and purely elective treatments, the differences between these two kinds of treatment have implications for whether patients are discharging their social responsibilities. If a treatment is not medically necessary but does create harms for others, including contributions toward health inequities in the destination country and public expenses for follow-up care in the patient's home country, then she can reasonably be held responsible for these negative effects. Such steps have been taken elsewhere. In Germany, for example, co-payments are required of patients needing treatment as a result of a "non-medically indicated measure such as cosmetic surgery, a tattoo, or a piercing" [67](*p.1188*). Even for medically-necessary treatments, however, any determination of whether the medical tourist has failed in her social responsibility will depend on that patient's ability to assess potential harms to the destination country and the degree of the patient's control over the decision to engage in the elective surgery.

If patients engaging in medical tourism do have a social responsibility to restrict their participation in this practice, we must be sure that this responsibility does not fall disproportionately on the poor, uninsured, and other vulnerable groups who may be driven into travel abroad for medical care due to a lack of options at home. The danger is that talk of patient responsibility can be used to further burden the most disadvantaged members of a community [83]. Any determination of whether a proposed social responsibility for medical tourists would unfairly burden certain patients will require reference to the particular context in which the patient acts, including whether her home health care system is public or private, the environmental health burdens faced by the patient, her socio-economic position within her community, and individual factors that might undermine her ability to access healthcare. While it is difficult to say in general and without reference to the particular circumstances of a patient what the extent of a medical tourist's social responsibility is, the claim that we have defended here is that individuals face a social responsibility to their health systems to use these systems efficiently and to protect fair access to others. By choosing to access the health systems of other countries, medical tourists expand the scope of this social responsibility, entailing new responsibilities to not unduly burden their home health systems and also to use the health systems of other countries both fairly and efficiently.

DISCUSSION

Voluntourism and medical tourism are both global health care practices that have dominant flows whereby citizens of the global north travel to the global south. As our discussion of both of these practices has shown, they each entail social responsibilities for their participants. By analyzing the similarities and points of divergence in the social responsibilities generated by voluntourism and medical tourism, we identify how our understanding of the social responsibilities of voluntourists can be illuminated by a discussion of the social responsibilities of medical tourists and vice versa. In Figure 1 we present a conceptual model that visualizes the similarities and points of divergence discussed in this section.

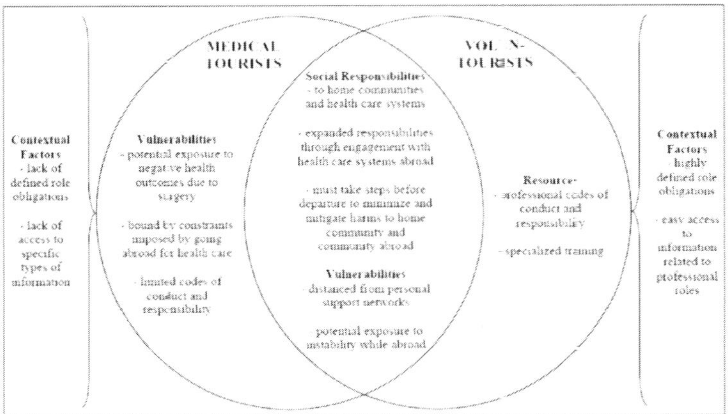

Figure 1. Overlaps and Dissimilarities in Medical Tourists' and Voluntourists' Social Responsibilities.

Overlaps

Physicians and patients both have social responsibilities toward their domestic communities and health care systems. Physicians have an obligation to ensure that local medical systems are equitable and accessible and do not create conditions that encourage medical travel. As we have noted, physicians are bound by professional codes of ethics that require them to serve the interests of those in need. Physicians are in a unique position to meet the medical needs of their communities, and to refuse to do so can serve to show a callous disregard for these needs. Patients, we have argued, have a social responsibility to use medical resources responsibly and to take steps to avoid worsening the health of those around them, including through the spread of infectious diseases. A patient who took no steps to protect the health of fellow community members would, through her actions, not demonstrate respect for their claim to having their basic health needs met.

While voluntourists and medical tourists have social responsibilities to the communities with which they choose to engage, they are also put into positions of vulnerability by engaging in these practices of global health care that are undertaken across vast distances. Both voluntourism and medical tourism may entail travel far from one's home community. This travel may create stresses, including separation from one's friends and family, cultural and linguistic differences, and anxiety during the time abroad [20, 22]. Voluntourists may face risks to their health and safety, particularly if they are traveling to a community that has poorly developed infrastructure, as will commonly be the case. Medical tourists are in a position of vulnerability as, like other patients, they face risks to their health from complications stemming from their medical procedures. But unlike most other patients, they often face these risks far from their support networks.

Persons engaging in voluntourism and medical tourism both can face exposure to political, social, and cultural instability. Voluntourists are called to administer care in communities abroad that are often impoverished, have poorly developed infrastructure, face political instability, and are exposed to endemic disease. While medical tourism is often advertised as providing patients with a safe and relaxing environment for care and recovery, they too can be exposed to unstable environments abroad. Many medical tourists were in Thailand during a recent outbreak of political instability, for example, and medical tourists may not be well informed about the local political conditions in the countries to which they are considering traveling [84]. Thus, both voluntourists and medical tourists, by choosing to travel abroad and engage in global health practices, are exposed to new vulnerabilities.

For both physicians and patients, the decision to travel to another country to receive or deliver health care serves to expand the range of the individual's social responsibility. The logic for this expansion of a pre-existing responsibility follows the rationale for the original social responsibility. Just as choosing to ignore the health needs of one's own community members when one could easily take steps to address these needs shows a disregard for others, engaging in voluntourism and medical tourism brings people into contact with new communities with their own distinctive needs. This contact creates new opportunities to take actions to meet local needs, or to ignore them altogether. Just as disregard for others' needs in one's original community would call into doubt one's commitment to others as having a right to adequate health, contact with a new community raises the possibility of similar, morally problematic inaction.

In order to ensure that they demonstrate concern for the needs of others and thereby discharge their social responsibilities, both voluntourists and medical tourists must take steps, before they travel abroad, to ensure that their choice to engage in these practices will not harm those with whom they come into contact. As we have observed, voluntourism raises the possibility of such harm if physicians fail to take into account the distinct needs of the local population, develop cultural and/or linguistic competency, or fail to ensure that the care they offer is sustainable. Medical tourists can encourage inequitable access to care in the countries to which they travel and may carry new infections to or create new costs for their home community. By taking steps to mitigate the potential for these harms prior to departure, voluntourists and medical tourists both help to discharge their social responsibilities.

Dissimilarities

While both voluntourists and medical tourists face new vulnerabilities in virtue of their decision to travel abroad, the types and degrees of vulnerabilities faced by each group will likely be different. The key difference in these vulnerabilities is linked to the roles that each group takes when travelling abroad. The medical tourist often enters into travel in a very vulnerable position as she is seeking care to address her health needs. While some forms of medical tourism for purely elective procedures like cosmetic surgery may not place the medical tourist in a position of great need, any medical procedure carries risks of adverse side effects and post-operative infections. Some procedures, like cardiac surgery, will place the medical tourist in a position of great vulnerability due to high risk of negative outcomes [85, 86].

While we should not discount the vulnerabilities faced by voluntourists, relative to medical tourists they will often be in a position of power due to the hierarchies implicit and explicit in the provision of medical care. The medical tourist may feel forced to travel abroad for care because of wait times for services or the high cost of medical care at home, particularly if the patient is uninsured [20]. The voluntourist, on the other hand, engages in this practice much more freely, though he or she may feel that doing so is part of an ethical obligation [7, 44]. The knowledge and position of voluntourists allows them actively to provide medical services and intervene in addressing the needs of others. By contrast, medical tourists seek medical care and may be bound by a range of geographical and cost constraints.

The role of physician is much better defined than that of patient. While we have argued that patients are a group to whom distinct social responsibilities are attached, they are a more loosely defined group with fewer clear norms of behaviour and less of a governing institutional structure. While we all are patients at some points in our lives, physicians make up distinct professions, the membership of which is shaped by professional bodies. These bodies can in part dictate which individuals can be counted as members and help set governing norms for their behaviour. Thus, the social responsibilities of physicians, including those who choose to act as voluntourists, are much better defined than those of patients, who lack professional bodies to develop codes of conduct. Those codes of patient responsibility that we have identified and discussed above are typically the result of public health care institutions choosing to set norms for their members. Many patients, particularly in privatized systems, will not fall under the umbrellas of these public bodies, however, and will not be as clearly governed by these norms. Moreover, by choosing to travel abroad for care and, as is typical, pay out-of-pocket for this care, medical tourists frequently opt out of public health care systems and thereby the norms that dictate their responsibilities. For these reasons, codes of social responsibility for medical tourists have been slower to develop than those for voluntourists.

Finally, we have suggested that both voluntourists and medical tourists have a social responsibility to eliminate or mitigate any risks of harm to others that may be a consequence of their decision to engage in these global health practices. As we have already observed, medical tourists' choices may be much more circumscribed than that of voluntourists. Moreover, the information available to the medical tourists, with which they may attempt to mitigate the risk of harms stemming from their actions, is much more limited. As medical tourists are often very sick and in pain, they may not have the energy or focus to try to bridge these informational gaps. This informational asymmetry is due, in part, to the training of physicians compared to that of the typical medical tourist. Most of these international patients will not have access to specialized medical knowledge and may not be aware of the potential for medical tourism to exacerbate health inequities in destination countries or contribute to the spread of infectious disease. While physicians engaging in voluntourism will frequently receive specialized training specific to the context of the community to which they will be traveling, medical tourists typically travel at their own volition and without any formal guidance. Medical tourists may travel with the assistance of medical tourism facilitators [29], but we have no evidence that these facilitators provide information to medical tourists that would be relevant to discharging their social responsibilities. Thus, relative to voluntourists, medical tourists will often find it very difficult to determine how to mitigate any negative consequences of their travel, if they are even aware that such risks exist.

Moving Forward

As we have discussed, physician voluntourism is seen a potentially ethically problematic approach to the provision of medical services in international settings, especially by students [37]. Hence, agencies that support medical student volunteers are beginning to insist on adequate pre-departure training to prepare them for the range of ethical issues they may encounter abroad [88]. Equipping volunteers for ethically responsible practices will require a transformative pedagogy [89], and the development of critical consciousness about the root causes of disparities in healthcare [90]. Pre-service medical training using international service-learning (ISL) opportunities appears to provide a promising experiential pedagogy for nurturing a sense of social responsibility and global citizenship among volunteers [24]. Unlike traditional voluntourism, ISL provides a platform for reciprocal, collaborative and mutual learning between a community and the volunteer. Volunteers are expected to develop a sense of critical awareness about the problems vulnerable communities face, and demonstrate ethical conduct and problem-solving skills as their experience in a given community unfolds. The focus of ISL is less on clinical skills development and more on developing an understanding of the social determinants of health that affect vulnerable communities. Interventions are designed in collaboration with communities in ways that are locally sustainable, enabling volunteers to learn how social determinants impact health and illness and health inequities [91].

Some training programs have successfully utilized the critical incident technique [92] to help physician volunteers explore how best to engage communities in ways that strive for social justice by understanding and acting to change the social structures that stifle individuals and communities due to unequal power relations, poverty and vulnerability. Physician volunteers are encouraged to develop a sense of professional and personal growth, and to examine critically what it means to be a socially responsible practitioner [93]. For example, many voluntourists seem to believe that being socially responsible means charity [60]. But charity can create dependency relationships whereas social responsibility aims at social justice, understood as developing sustainable relationships based on mutual respect. It involves working with and for communities to enable what they feel is best for them rather than using a paternalistic approach. Dickson and Dickson [60], identify a list of personal attributes that physicians need to develop as part of their professionalization and to act responsibly that include: a concern with global equity; a commitment to redressing injustices in healthcare; respect for diversity; openness to mutual learning; and embracing ethical values like human rights and social justice. The professionalization of physicians gives them norms by which their social responsibilities as voluntourists are increasingly clearly stated. It also gives physicians the information and expertise with which they may act on these norms.

By comparison, medical tourists are given little guidance on their social responsibilities and little capacity to act on these norms. In order to address this gap, there is great need for the development of guidelines for medical tourists on how they can prepare for their travel, engage in this practice with a minimum of personal risk, and take steps to maintain their own health and arrange follow-up care after travel. Because of the fragmented nature of the many community and national systems for distributing health care, these guidelines will likely take time to achieve widespread uptake. A starting point would be to have specific countries develop model patient guidelines that can be adapted and then adopted by other countries, thus encouraging better patient protections through their examples. These guidelines could be distributed by medical tourism facilitators or travel health providers.

The latter group could be trained to give other advice on ethical medical tourism to their patients (Eyal: Global Health Impact Labels, submitted).

International patients need transparency so that they can make informed and responsible health choices. Given the lack of regulation in the medical tourism industry - due in part to its newness and global nature - patients may have difficulty getting information on the practices of the medical facilities they wish to travel to and may have no way of judging the accuracy of the information that they do receive [29]. One way of addressing this problem would be to develop ethical buying guidelines for patients engaging in medical tourism akin to those for consumers of other products like apparel, coffee, and chocolate. Patients will face difficulties in developing these guidelines on their own given the informational gaps discussed above and the fact that these patients may be in pain or short of time given their ill health. Moreover, the medical tourism industry may be reluctant or unable to develop these guides and to self-regulate both because of a reluctance to place restrictions on their business and because of the fragmentary nature of this industry at this early stage in its development. Non-governmental groups can take on the role of regulators, developing information from medical tourism facilities, assessing the accuracy of this information, and rating facilities for their tendency to promote the positive and mitigate the negative effects of medical tourism. While a non-governmental agency will not be able to compel medical tourism facilities to participate in a rating scheme, patients who see engaging in medical tourism responsibly as a moral obligation could create market pressure for participation (Eyal: Global Health Impact Labels, submitted). Alternately, patients' home country governments may need to take a more active role in regulating the medical tourism industry and providing patients with more information and transparency [94]. As medical tourism facilitators and other industry members can escape restrictions on their practices if they are developed piecemeal, there is a strong argument for finding ways to increase the information available to medical tourists and raising awareness of the potential negative effects of this practice. Doing so will allow medical tourists to apply pressure within the market for medical tourism services to counter problematic elements within this industry (Eyal: Global Health Impact Labels, submitted).

Medical professionals can and should help to develop the norms and infrastructure needed for medical tourists to discharge their social responsibilities. Doing so can be connected to these physicians' own responsibilities, as medical tourism will increasingly affect patients' home countries through the need for expensive follow-up care and the need to create systems to monitor and minimize the spread of infectious disease. Moreover, physicians can be implicated in the push factors encouraging medical tourism [20]. If patients are traveling abroad due to a lack of insurance, perceived lengthy wait times for care, or the unavailability of certain procedures, then health professionals must ask what their role is in reducing these factors and caring for the patients who choose to seek care abroad. While medical tourism and voluntourism are distinct global health practices, those engaging in them are closely connected through the increasingly global nature of health care, travel, and the business of medicine.

SUMMARY

Voluntourism and medical tourism share many qualities, including being tied to an expansion of social responsibilities due to an individual's choice to engage in a global health practice. Both voluntourists and medical tourists have a social responsibility to limit the risk of harms to members of their home and destination countries and to take steps to ensure that the global health practices in which they engage allow for sustainable development in their destinations. The complications associated with health care in low and middle income countries mean that voluntourists and medical tourists must prepare for their travel in order to avoid inadvertent harms to others. Yet, as we have seen, physician volunteers are much better prepared to do so than medical tourists in virtue of their membership in a well-organized professional group with a strong historical sense of social responsibility.

While physicians must continue developing and enforcing guidelines for discharging their social responsibility while practicing abroad, the field of medical tourism is less well prepared to develop the tools and guidelines needed for socially responsible medical tourism. As medical tourists themselves are not a well-organized group and may not be aware of the implications of their choice to engage in medical tourism, it is important that better organized and informed groups help to fill this vacuum. Physicians and other health professionals are members of groups that can draw on their knowledge, skills, and sense of social responsibility to help develop guidelines for responsible medical tourism. But it will also be important for medical tourism industry groups to engage in this process, including professional organizations like the Medical Tourism Association and medical tourism facilitators

[95]. Fortunately, there is a long history of professional and business groups developing guidelines for socially responsible practice, from which medical tourism industry groups can learn [96]. It will be up for the rest of society, however, to help guide these groups and to ensure that they are encouraged or even required to develop these guidelines.

ACKNOWLEDGEMENTS AND FUNDING

Funding was provided by a Catalyst Grant from the Canadian Institutes of Health Research. Thanks also go to Krystyna Adams for research and referencing help.

COMPETING INTERESTS

The authors declare that they have no competing interests.

AUTHORS' CONTRIBUTIONS

JS wrote the introduction and summary, parts of the background, and the medical tourism sections of the discussion and edited throughout. SD wrote the social responsibility and voluntourism sections of the background and discussion. VC contributed to the background section, and contributed greatly to conceptualization and editing of this manuscript. All authors provided feedback on drafting this paper and approved the final version of the manuscript.

REFERENCES

1. Bowen HR: Social responsibilities of the businessman. 1953, Harper

2. Freeman I, Hasnaoui A: The Meaning of Corporate Social Responsibility: The Vision of Four Nations. Journal of Business Ethics. 2010

3. Ginsburg RB: Access to Justice: The Social Responsibility of Lawyers. Journal of Law & Policy. 2001, 7: 1-

4. Medcalf M: Advancing Social Justice Through an Interdisciplinary Approach to Clinical Legal Education: The Case of Legal Assistance of Windsor Rose Voyvodic. Journal of Law & Policy. 2004, 14: 101-

5. Foreman S: Social responsibility and the academic medical center: building community-based systems for the nation's health. Academic Medicine. 1994, 69: 97.

6. Faulkner LR, McCurdy RL: Teaching medical students social responsibility: the right thing to do. Academic Medicine. 2000, 75: 346-

7. Weed DL, McKeown RE: Science and social responsibility in public health. Environ Health Perspect. 2003, 11: 1804-1808.

8. Garriga E, Melé D: Corporate Social Responsibility Theories: Mapping the Territory. Journal of Business Ethics. 2004, 53: 51-71.

9. Luban D: Lawyers and justice. 1988, Princeton: Princeton University Press

10. Bloche MG: Clinical Loyalties and the Social Purposes of Medicine. JAMA. 1999, 281: 268-274.

11. Crane A, McWilliams A, Matten D, Moon J, Siegel D: The Oxford Handbook of Corporate Social Responsibility. 2008, Oxford: Oxford University Press

12. Ite UE: Multinationals and corporate social responsibility in developing countries: a case study of Nigeria. Corporate Social Responsibility and Environmental Management. 2004, 11: 1-11.

13. Bowie NE, Bowie A: Business ethics: A Kantian perspective. 1999, Blackwell Publishers

14. Reed C: Medical Tourism. CDC Health Information for International Travel. Edited by: Brunette G, Kozarsky P, Magill A, Shlim D. 2010, Philadelphia: Elsevier

15. Kangas B: Complicating Common Ideas about Medical Tourism: Gender, Class, and Globality in Yemenis' International Medical Travel. Signs. 2011, 36: 327.

16. Wibulpolprasert S, Pachanee C, Pitayarangsarit S, Hempisut P: International service trade and its implications for human resources for health: a case study of Thailand. Human Res Health. 2004, 2: 10.

17. Laugesen MJ, Vargas-Bustamente A: A patient mobility framework that travels: European and United States-Mexican comparisons. Health Policy. 2010, 97: 225-231.

18. Turner L: "Medical tourism" and the global marketplace in health service: U.S. patients, international hospitals, and the search for affordable health care. Int J Health Serv. 2010, 40: 443-467.

19. Shah R: The international migration of health workers: ethics, rights and justice. 2010, Basingstoke: Palgrave Macmillan

20. Crooks VA, Kingsbury P, Snyder J, Johnston R: What is known about the patient's experience of medical tourism? A scoping review. BMC Health Services Research. 2010, 10: 266.

21. World Health Organization: The global shortage of physicians and its impact. 2007, Geneva

22. Pezzella AT: Volunteerism and humanitarian efforts in surgery. Curr Probl Surg. 2006, 43: 848-929.

23. Association of American Medical Colleges: AAMC Medical Student Questionnaire: Summary Report. 2006

24. Dharamsi S, Richards M, Louie D, Murray D, Berland A, Whitfield M, Scott I: Enhancing medical students' conceptions of the CanMEDS Health Advocate Role through international service-learning and critical reflection: A phenomenological study. Med Teacher. 2010, 32: 977-982.

25. Isaacson G, Drum ET, Cohen MS: Surgical missions to developing countries: Ethical conflicts. Otolaryngol Head Neck Surg. 2010, 143: 476-479.

26. Morgan MA: Another view of "humanitarian ventures" and "fistula tourism". Int Urogynecol J Pelvic Floor Dysfunct. 2007, 18: 705-707.

27. Bezruchka S: Medical tourism as medical harm to the Third World: Why? For whom?. Wilderness Environ Med. 2000, 11: 77-8.

28. Vastag B: Volunteers see the world and help its people. JAMA. 2002, 288: 559-565.

29. Johnston R, Crooks VA, Snyder J, Kingsbury P: What is known about the effects of medical tourism in destination and departure countries? A scoping review. Int J Equity Health. 2010, 9: 24.

30. Hopkins L, Labonté R, Runnels V, Packer C: Medical tourism today: What is the state of existing knowledge?. J Pub Health Policy. 2010, 31: 185-198.

31. Lunt N, Carrera P: Medical tourism: assessing the evidence on treatment abroad. Maturitas: An international journal of mid-life health. 2010, 66: 27-32.**V**

32. Shetty P: Medical tourism booms in India, but at what cost?. Lancet. 2010, 376: 671-72.

33. Chanda R: Trade in Health Services. Trade in Health Services: Global, Regional and Country Perspective. Edited by: Drager N. 2002, Washington DC: Pan-American Health Organization

34. Woodward ND, Beaglehole R, Lipson D: Globalization, public goods and health. Trade in Health Services: Global, Regional and Country Perspectives. Edited by: Drager N. 2002, Washington DC: Pan-American Health Organization

35. De Arellano R: Patients Without Borders: The Emergence of Medical Tourism. Int J Health Serv. 2007, 37: 193-198.

36. Bookman MZ, Bookman KR: Medical Tourism in Developing Countries. 2007, New York: Palgrave MacMillan

37. Thomas G, Krishnan S: Effective public-private partnership in healthcare: Apollo as a cautionary tale. Indian J Med Ethics. 2010, 7: 2-4.

38. Kenny N, Shelton W: Volume 10: Lost Virtues. Advances in Bioethics. 2006, 10: 1-233.

39. Twohig PL, MacDonald C: Professionalism and the social role of medicine. Am J Bioethics. 2004, 4: 3-5.

40. Schroeder SA: Training an appropriate mix of physicians to meet the nation's needs. Acad Med. 1993, 68: 118-122.

41. American Medical Association: Original Code of Ethics. 1847

42. Boelen C: Building a socially accountable health professions school: towards unity for health. Educ Health (Abingdon). 2004, 17: 223-231.

43. Welie JV: Is dentistry a profession? Part 1. Professionalism defined. J Can Dent Assoc. 2004, 70: 529-532.

44. Pellegrino ED: Professionalism, Profession and the Virtues of the Good Physician. The Mount Sinai Journal of Medicine. 2002, 69: 378-384.

45. Freidson E: Profession of medicine: a study of the sociology of applied knowledge. 1988, Chicago: University of Chicago Press

46. Mirko D: Western medical thought from antiquity to the Middle Ages. 1998, Cambridge: Harvard University Press

47. Humphreys R: Sin, organized charity and the poor law in Victorial England. 1995, New York: St. Martin's Press, In

48. Whitehead M: The concepts and principles of equity and health. Health Promot Int. 1991, 6: 217-228. 10.1093/heapro/6.3.217.

49. Verma S: Honouring the social contract: medical schools take social responsibility seriously. University of Toronto Bulletin. 2005

50. Boelen C, Woollard B: Social accountability and accreditation: a new frontier for educational institutions. Med Educ. 2009, 43: 887-894.

51. Boelen C, Heck J: Defining and measuring the social accountability of medical schools. 1995, Geneva: World Health Organization

52. Maki J, Qualls M, White B, Kleefield S, Crone R: Health impact assessment and short-term medical missions: a methods study to evaluate quality of care. BMC Health Serv Res. 2008, 2: 121.

53. Association of American Medical Colleges: GQ medical school graduation questionnaire: all schools summary report. 2009, Washington, DC

54. Green T, Green H, Scandlyn J, Kestler A: Perceptions of short-term medical volunteer work: a qualitative study in Guatemala. Globalization and Health. 2009, 5: 4.

55. Sichel B: "I've come to help": Can tourism and altruism mix?. Briarpatch Magazine. 2006

56. Roberts M: Duffle Bag Medicine. JAMA. 2006, 295: 1491-1492.

57. Baraldi J: A harm in 'medical tourism': The poor need lasting efforts to improve global health, not feel-good field trips. The Philadelphia Inquirer. 2009

58. Shah S, Wu T: The medical student global health experience: Professionalism and ethical implications. J Med Ethics. 2008, 34: 375-378.

59. Bhat S: Ethical coherency when medical students work abroad. Lancet. 2008, 372: 1133-1134.

60. Dickson M, Dickson G: Volunteering: Beyond and act of charity. J Can Dental Assoc. 2005, 71: 865-869.

61. Crooks VA, Andrews G, (Eds): Primary Health Care: People, Practice, Place. 2009, Aldershot and Vermont: Ashgate Publishers

62. Cueto M: The origins of primary health care and selective primary health care. Am J Pub Health. 2004, 94: 1864-1874.

63. Kirkness VJ, Barnhardt R: First Nations and Higher Education: The Four R's - Respect, Relevance, Reciprocity, Responsibility. Knowledge Across Cultures: A Contribution to Dialogue Among Civilizations. 2001, Hong Kong: Comparative Education Research Centre, The University of Hong Kong

64. Wikler D: Personal and Social Responsibility for Health. Ethics & Int Affairs. 2002, 16: 47-55.

65. Schmidt H: Just Health Responsibility. J Med Ethics. 2009, 35: 21-26.

66. Buyx A: Personal Responsibility for Health As a Rationing Criterion: Why We Don't Like It and Why Maybe We Should. J Med Ethics. 2008, 34: 871-874.

67. Schmidt H: Patient's Charters and Health Responsibilities. BMJ. 2007, 335: 1187-1189.

68. Commission on the Future of Health Care in Canada: Building on Values: The Future of Health Care in Canada. 2002, Saskatoon: Commission on the Future of Health Care in Canada

69. Scotland NHS: The NHS and You. 2006, Scotland

70. Virginia Department of Health and Human Resources: West Virginia Medicaid Member Agreement. 2006, Virginia

71. Cabinet for health Cabinet for Health and Family Services: KyHealth Choices: Kentucky's Medicaid Transformation Initiative. 2006, Kentucky

72. Steinbrook R: Imposing Personal Responsibility for Health. New Enlg J Med. 2006, 355: 753-756.

73. Bishop G, Brodkey A: Personal responsibility and physician responsibility--West Virginia's Medicaid plan. N Engl J Med. 2006, 355: 756-758.

74. Pennings G: Ethics without Boundaries: Medical Tourism. Principles of Health Care Ethics. Edited by: Ashcroft RE, Dawson A, Draper H, McMillan JR. 2007, Sussex: Wiley & Sons Ltd, 2

75. Turner L: From Durham to Delhi: "Medical Tourism" and the Global Economy. Comparative Program on Health and Society Lupina Foundation Working Papers Series 2006-2007. Edited by: Cohen-Kohler JC, Seaton MB. 2007, Toronto: University of Toronto Press, 109-131.

76. Turner L: Canadian Medicare and the Global Health Care Bazaar. Policy Options. 2007, 73-77.

77. Carabello L: A Medical Tourism Primer for U.S. Physicians. J Med Practice Manag. 2008, 23: 291-

78. Unti JA: Medical and surgical tourism: the new world of health care globalization and what it means for the practicing surgeon. Bull Am Coll Surg. 2009, 94: 18-25.

79. Snyder J, Crooks VA: Medical Tourism and Bariatric Surgery: More Moral Challenges. Am J Bioethics. 10: 28-30.

80. Kumarasamy KK, Toleman MA, Walsh TR, Bagaria J, Butt F, Balakrishnan R: Emergence of a new antibiotic resistant mechanism in India, Pakistan, and the UK: a molecular, biological, and epidemiological study. The Lancet Infect Dis. 2010, 10: 597-602.

81. Keckley PH, Underwood HR: Medical Tourism: Update and Implications. 2008, Washington, D.C.: Deloitte Centre for Health Solutions

82. Burkett L: Medical Tourism: Concerns, Benefits and the American Legal Perspective. J Legal Med. 2007, 28: 223-245.

83. Schmidt H, Voigt K, Wikler D: Carrots, Sticks, and Healthcare Reform - Problems with Wellness Incintives. New Engl J Med. 2010, 362: 3

84. Einhorn B: Medical Tourism: Surviving the Global Recession. BusinessWeek: Asia. 2008

85. Grace M: State of the heart: a medical tourist's true story of lifesaving surgery in India. 2007, Oakland CA: New Harbinger Publications

86. Turner L: 'First World Health Care at Third World Prices': Globalization, Bioethics and Medical Tourism. BioSocieties. 2007, 2: 303-325.

87. White MT, Cauley KL: A caution against medical student tourism. J Am Med Assoc Ethics. 2006, 8: 851-854.

88. Drain PK, Holmes KK, Skeff KM, Hall TL, Gardner P: Global health training and international clinical rotations during residency: current status, needs, and opportunities. Acad Med. 2009, 84: 320-325.

89. Mezirow J: Transformative Dimensions of Adult Learning. 1991, San Francisco: Jossey-Bass

90. Kumagai AK, Lypson ML: Beyond cultural competence: critical consciousness, social justice, and multicultural education. Acad Med. 2009, 84: 782-787.

91. Spiegel JM, Dharamsi S, Wasan KM, Yassi A, Singer B, Hotez PJ, Hanson C, Bundy DA: Which new approaches to tackling neglected tropical diseases show promise?. PLoS Med. 2010, 18: 7-

92. Flanagan JC: The critical incident technique. Psycholog Bull. 1954, 51: 327-358.

93. Mofidi M, Strauss R, Pitner LL, Sandler ES: Dental students' reflections on their community-based experiences: the use of critical incidents. J Dent Educ. 2003, 67: 515-23.

94. Public Health Agency of Canada: Statement on Ethics and Travel. 2003, Canada Communicable Disease Report, 29:

95. Turner LG: Quality in health care and globalization of health services: accreditation and regulatory oversight of medical tourism companies. Int J Quality in Health Care. 2010, 23: 1-7. 10.1093/intqhc/mzq078.

96. Carroll AB: Corporate Social Responsibility. Bus & Society. 1999, 38: 268-295.

11

CHAPTER

Exploring Health Care and Medical Tourism in a Modernizing Society: Journey in Chennai, India

Janani Krishnaswami,

Internal Medicine and Preventive Medicine at Kaiser Hospitals-San Francisco and the University of California-San Francisco

THE START OF A JOURNEY

I am riding in the front seat of a tiny Ford Icon, taking in the left-sided passenger view as we bump along narrow dirt roads. The driver is cheerfully unperturbed as he navigates various obstacles in our path: children, motorists, stray dogs, cycle rickshaws, carts of bright red tomatoes. By the end of my trip, my mind—and stomach—will be desensitized to the labyrinth of vehicles, pedestrians, and animal life constituting traffic in Chennai. But today is my second day; thus, I still shut my eyes in instinctive terror as our car swerves to avoid the mammoth trucks lurching toward us.

Chennai, India: A city in which I have never lived, but whose ancient culture shapes my roots. My memories of this south Indian city centered on its role as default destination of onerous family vacations. Back then, the sweltering summer weeks of bumpy rickshaw rides, obligatory extended-family visits and inevitable digestive woes felt more like forced medicine than my parents' claim of "relaxation." Yet these trips nurtured my subconscious appreciation for Indian culture, as I studied Sanskrit literature and trained as a student of Indian classical vocal music. Over the years, summer trips lessened in frequency, but my connection to Chennai grew tenacious as I pursued its art forms.

That connection had planted the seed for this trip. As an almost-physician, I was returning to Chennai with two specific goals: to provide medical service, while immersing myself in the culture inspiring my cherished hobbies.

Since my last visit (nearly a decade earlier), both of us had changed rather substantially. I'd moved across continents, worked for a behemoth investment bank and a tiny nonprofit agency, studied economics in England and parasitology in Texas—and now, months away from medical school graduation, was embarking on a future career in preventive medicine and public health. In India, transformation permeated society. A patriotic, protradition movement had sparked a nationwide flurry of renaming Indian cities in indigenous language (Bombay to Mumbai, Calcutta to Kolkata, Madras to Chennai) even while popular media captured evidence of India's resolute modernization: humble tea stalls replaced by Internet cafes, expansive rice fields now home to gleaming "tech parks," mobile *dosa-poori-masala* stands juxtaposed near freshly painted Pizza Huts, bright yellow auto-rickshaws lost in the roar of sporty Honda Citis.

And so it was thus, on this occasion of visiting a home I secretly feared would seem rudely foreign, that I found myself cowering in the passenger side of a Ford Icon.

A SNAPSHOT OF DAILY LIFE IN CHENNAI

Chennai operates in a chaotic, stubbornly functional context. First, people are everywhere. Chennai's 7 million people share a space of 180 square km, representing a population density of 24,231 people per square kilometer. (Los Angeles, one of the most population-dense cities in the US, has a density of 3170 people/km^2).[1-4]

Life in Chennai splashes boisterous color on the seemingly mundane. On the road, trucks groaning with loads of cargo are festooned with painted curlicues and canary-yellow paint jobs; shifting into reverse gear, they emit tinny renditions of various *bhajans*, Hindi film songs, or the national anthem. On nearly every corner, stores sell milk and pistachio *("pistha")* biscuits along with artists' paintbrushes, crochet kits, an array of sketchbooks, and Fevicol-brand craft glue. Street vendors wheeling carts of fresh vegetables, sugarcane juice and customers' folded laundry advertise their wares by shouting at the top of their lungs, creating a collective vocal cacophony that only a professional mother-in-law cooking in apartments above can interpret. Everyone loves music, and fittingly, every street in Chennai pulses to a perpetual soundtrack: Bollywood, classical, *bhajans*, instrumental mandolin/veena/violin, drum beats of *tabla* and *mridangam*. And every December, the whole city celebrates the Chennai Music Season, a five-week festival featuring hundreds of heavily attended classical music and dance concerts, lectures, demonstrations, and performances.[5]

Religious diversity is often matter-of-fact on Chennai's streets, even as India's borders are caught in religious quagmire. Travel down one road and you will see a prestigious convent school facing a billboard that proclaims *Jesus Loves You*, catercorner to a procession celebrating *Ganesha Chaturthi* at a Hindu temple. On another corner, traditional Muslim prayer call echoes from a mosque tucked in between "Krishna Tailoring Stores" and "Muhammed and Sons Jewelers." Turbans, *burqas, saris*, suits, *dhotis*, dresses, *shalwar kameez*, pants and shirts ("pant-shirt" in Chennai-speak), and sometimes no shirts are all accepted Chennai attire. It can sometimes seem as though the city is in a state of perennial celebration of some festival or holiday, with the sight of a sequin-bedecked, palanquin-hoisting, hymn-chanting and dancing group in the street almost as commonplace as rush-hour traffic.

A Bit Lost

The address I've been given is "19, 12 Cross Street, Indira Nagar, Chennai." The car meanders through increasingly pot-holed, nameless dirt roads; I wonder restlessly how we know where we're going. "Oh, 12 Cross Street," my driver, Shekhar, had said confidently. "We'll get there, we'll get there. No problem." Shekhar, of course, never said we *couldn't* do anything. Drive across town in 20 minutes during Chennai's rush hour, which makes Rockefeller Plaza at Christmastime look like Germany's Autobahn? "Of course, we'll get there, we'll get there. No problem." Everything is possible for Shekhar.

Except, apparently, finding this clinic. The car slows to a pathognomonic I'm-lost pace, and I peer worriedly out the window. Still no street signs. Flat-roofed cement houses—pastel pink, lime yellow, mint green—line the road, each enveloped by lush green creepers and vibrant hibiscus and jasmine. In lieu of door numbers, each house features a name painted in neat white letters atop an iron-railing fence. *Lakshmi Nivas. Swaathi. Sai Krupa*. Despite this decidedly non-numerical pattern, we creep by in hope of finding number 19.

The car pulls to a stop in front of a turbaned man idly chewing a betel leaf. Shekhar rolls down the window and beckons. *"Are, 12 Cross Street engu irruku theriyama?"* [Hey, do you know where 12 Cross Street is?]

he calls. The man spits out the leaf, squints for a moment and points. *"Angu* left *thirumbi, conju neru pettu, inooru ora* right..." [Turn left there, then go straight for a while, then another right ...]. I try to ignore the angry red betel stains on his teeth and instead concentrate on trying to understand the circuitous route. I'm lost after the first "left" and "right," but Shekhar nods in apparent excellent understanding. And we're off once again, narrowly dodging the cow that suddenly is in front of us, placidly turning its tail.

SURVEYING CHENNAI'S HEALTH

The context of health in Chennai is framed by a mammoth, decade-long rollercoaster ride of growth and development.

As India expanded its information technology (IT) sector and liberalized foreign investment laws, wealth flooded into Chennai's economy. Chennai now serves as one of India's major IT and IT-related services exporter, with its newly constructed "IT Corridor" employing over 300,000 people. Upward mobility thus touches and transforms the lives of many once-poor Chennai residents.[6–9] India's poverty rate declined over 50% from 1983 to 2007, and the 2006 South Asian Economic Report[10] noted a "remarkable" reduction in the rate of absolute poverty in the region surveyed "primarily due to accelerated growth in India." Among Indian states, Tamil Nadu ranks in the top five in terms of level of urbanization and literacy rate and has a lower-than-average poverty ratio. And in the 2009 Global Rankings Quality of Life Survey—covering 254 cities and conducted by the international human resources firm ECA International—Chennai was the top-ranked Indian city in which to live.[11–14]

I do give allopathic treatment for the pain and swelling. But that cannot change lives. A daily yoga practice, however, can— it improves flexibility, lean muscle mass, peace of mind all in one ...

Health care and health delivery are important beneficiaries of India's economic boom. The preceding decade's newfound wealth spurred grassroots efforts and nongovernmental organizations to fill basic public health needs: waste disposal, hygiene and sanitation, and access to clean water. Tangible examples include Friends of the Beach, an initiative in 1999 that installed garbage bins and public toilet facilities on Chennai's heavily polluted beaches and provides stable wages to slum-dwellers in exchange for beach cleaning and construction services.[15] The World Bank-supported Tamil Nadu Health Systems Project is a hospital-centered public health effort; this five-year, $110 million project will roll out electronic information systems in the state's district hospitals, streamline disposal of hospital waste and construct 24-hour OB/GYN centers.[16,17] And in the realm of drinking water access, the successful completion of the Sai Ganga Water Project represented a milestone. This multimillion-dollar initiative provides safe drinking water to millions of Tamil Nadu citizens, including the very poorest communities and slum dwellers, via groundwater pipeline connections from the neighboring water-rich state of Andhra Pradesh.[18]

Such measures dent the devastating death toll wreaked by India's endemic infectious agents. Tamil Nadu achieved a near-100% target vaccination rate from 2007-2008, and the death toll from six major vaccine-preventable diseases is zero.[13,19] Even once-rampant, "classic" developing world conditions are declining: in Chennai, malaria's incidence fell 43% between 2001–2008, and leprosy's prevalence is currently less than 1/10,000 in Tamil Nadu.[20,21] Expanding health care infrastructure also spurs dramatic increases in medical treatment and diagnosis. In 2008, Tamil Nadu's patients obtained 19.4 million lab tests, 19,000 Pap smears, and 600,000 surgeries.[22]

Despite such advances, however, poverty and poor health still trouble many residents of Chennai. Almost ten times as many infants die at birth in Chennai compared to Beijing; the average Chennai female can expect to live only about 67 years. The city's residents are served by about 1800 physicians, leading to a ratio of 42 physicians per 100,000. (In the US, the average is 286 physicians/100,000).[23–25] Even as enterprising students graduate from Chennai's institutes of higher education, others drop out of poorly funded, deplorably run government schools where student *and* teacher attendance is optional, desks and chairs are a luxury, and basic infrastructure is in flux. For example, a recent study found 38% of these schools lacked proper toilet facilities for its students. The city's overall literacy rate hovers at 73-78% (soaring from 46.7% in 1991), but literacy rates lag behind nearly 10-15% among poor classes, where sanitation, air pollution and improper waste disposal hinders quality of life.[26–30]

The dichotomy between rich and poor is stark in gentrified locales like Besant Nagar. Here, idyllic gardens and glass-doored shopping malls featuring "export-only" clothing and goods cater to the area's upper-middle class. One corner turn away on the paved road to the beach, however, is Oduma Nagar, a fishing colony and slum. In the slums of

Chennai, populations exceeding the size of small European countries coexist on a few square kilometers. For slum inhabitants, mundane needs and activities—bathing, garbage disposal, drinking water, and defecation—are daily battles. Garbage and human waste invariably ends up in the same water used for bathing and drinking, promulgating a vicious cycle of parasitic and communicable disease.[31–33]

ENTERING THE CLINIC

A khaki-uniformed security guard paces pensively by the wrought-iron gate encircling the cream-colored house. He pauses as we pull to a stop. *"Idhu* doctor *veettu ah*, doctor *Krishna Raman?"* [Is this Dr Krishna Raman's house?] Shekhar questions the guard. I peer anxiously as the guard smiles and nods vigorously, briefly pondering the irony that the word for "doctor" in vernacular Tamil—a hybrid of Tamil and English—is "doctor." Shekhar turns to me and nods with vigor and triumph; with good reason, as against all seeming odds, he's successfully brought me to my destination. Although, I fret, this so-called "clinic" looks much more like a house. Then I see the small, thatched-roof annex tucked alongside the house, bearing a hand-painted sign and a red cross inside a circle: "Dr Krishna Raman—Medical Doctor."

I gingerly step down from the car, my relatively pampered muscles bruised from the jarring hour of swerving over Chennai's obstacle paths. Then the reality of the moment floods over me. Here it is: the clinic of Krishna Raman, MBBS, FCCP, graduate of the BKS Iyengar Institute.[34]

Traveling down the inevitable extended-relative grapevine earlier, news of my medical "rotation" with Dr Raman had sparked much excitement. "Just one look at the frozen shoulder, and he knew exactly what to do: a few medicines, creams, and some yoga postures and I was completely back to normal," my aunt had gushed. My grandmother had solemnly agreed. "Dr Krishna Raman, he's a big name," she said with hushed reverence. "He's been on TV, helped many people ... you're lucky to be able to work with him." Although I have grown accustomed to my relatives' proclivity for exaggeration (eg, "air-conditioned" referring to "extra ceiling fan"), I figured at least some truth inspired this exuberant praise. A perusal of the clinic's Web site and Dr Raman's publications introduced me to his mission of providing quality health care to all patients regardless of socioeconomic status. And his treatment came with an added twist: allopathic medicine supplemented with an emphasis on lifestyle changes, including the incorporation of a personal practice of yoga.

Brushing beads of sweat off my forehead, I leave my slippers outside the door—the custom in Chennai—and step barefoot onto the mosaic-tiled floor of the clinic. The room is empty, except for the clamoring rattle of a fan, which seems to be producing more noise than breeze. A bare light bulb flickers overhead, casting feeble light over a wooden table placed in the center of the room. Five chairs are neatly lined in a row, empty except for a stack of magazines and newspapers: *Time. India Today. The Hindu*.

"Welcome!" I am greeted by a trim, elderly gentleman wearing a light green polo shirt and cleanly pressed khaki pants. Dr Raman ushers me inside to give me the one-room clinic tour. It's short: there's the waiting room, which I've already seen, and the combined exam-room/physician's office featuring familiar fixtures of outpatient medicine—a Dell computer station, exam table and the less familiar x-ray viewing box.

Dr Raman shows me the day's schedule, explaining that he routinely sees between 40–50 patients a day. Most come in with a chief complaint of musculoskeletal pain. "There are sometimes significant social issues," he says in lightly accented English. "Some women come with carpal tunnel syndrome from rolling *chapattis* all day for 20 in-laws, for example. It's impractical to tell them to stop cooking, when it's what everyone—including the patient—expects." I nod, wondering how anyone could tolerate standing even slightly near heat-generating kitchen equipment in the searing 100-degree-plus weather, especially in 9 yards of a silk fabric sari. "But I get patients from all backgrounds," he continues, leaning back in his chair as I take a seat in front of the desk. "Muslim, Hindu, Sikh, Christian, Tamil, Hindi, Telugu, Marwari. There are those who can't afford to pay—the care is free for them. Others are successful businessmen and women; they pay the regular fees." (Regular fees, I find out, are anywhere from 250—800 rupees [INR], $6-$12.)

Dr Raman also eagerly expounds on the aspect of yoga and lifestyle education interweaved into his clinic services. "These days, the nine-to-five tech jobs ... so many people are sitting, no such walking or activity as they had in the past. Only a decade ago, people walked more. Activity was built into their lives: walk three miles to school, to go to the temple and to do 40 *pradakshinams*.[a] Now it's staring at the computer screen."

"So naturally," he continues, "muscle pain complaints are the biggest part of my practice. Herniated disks, osteoarthritis, carpal tunnel syndrome, chondromalacia. A lot of it, of course, is sedentary lifestyle and being overweight. For these people, I do give allopathic treatment for the pain and swelling. But that cannot change lives. A daily yoga practice, however, can—it improves flexibility, lean muscle mass, peace of mind all in one," he concludes triumphantly.

I had half-expected an introductory discourse on Chennai's clinical epidemiology covering topics such as Pott's spine, leishmaniasis, and leprosy; my mind now shifts gears in the face of a medicine demographic echoing my patient encounters in the US. Dr Raman turns to his computer and, with a few mouse clicks, brings me even closer to home as he pulls up the ubiquitous blue screen of a Microsoft PowerPoint presentation. "Here, you can take a look at some MRIs and x-rays of some of my patients," he explains as we click through slides containing radiologic proof of patients' suffering and photos of patients (without faces blackened out as in my lectures back home) trying out yoga. There are sari-clad women holding onto a wall and twisting their spine; somewhat pot-bellied men stretching up valiantly to the ceiling; "before" and "after" MRIs that show the progression of a herniated disc back into its rightful place after a year of yoga. I'm impressed.

Dr Raman looks up abruptly at the clock. "Well, it's 8:30. Let's see our patients."

The Rise of Chennai's Medical Tourism Industry

As India tackles basic deficiencies in public health indicators, another phenomenon transpires in parallel: the explosion of an elite private health care sector worth nearly $15 billion. Catering to the upper-class segments of Indian society with "five-star" hospitals, this emerging industry is responsible for India's "extreme makeover" of health care.[35]

Each year, over 150,000 patients fly in from around the world to receive comparatively lower-price treatment or escape long waiting lists. Their stories are featured in op-ed pieces and magazines: Bill W from California, denied health insurance because of a high PSA and suspected cancer, paying cash to receive a transurethral resection of the prostate (TURP) in New Delhi's Fortis Hospital for a quarter of the cost back home; Mohammed S from Kuwait, undergoing an eight-hour removal of a glioblastoma in Chennai's MIOT hospital only a few days after receiving his diagnosis at home.[35,36]

The medical tourism industry in India traces its roots to a Chennai hospital .... which opened its first international branch in 2007.

Nearly 50% of patients treated in Chennai travel from outside Tamil Nadu. Chennai's hospitals go to great lengths to attract this lucrative group of patients, offering city tours, swanky hotel rooms for guests, airport pick-up and drop-off, and hours of one-on-one time with staff physicians. All with calculated reason, for the industry is forecasted to grow at a rate of 13% per year for the next six years, eventually constituting 3% to 5% of the health care delivery market and contributing $1-$3 billion additional revenue for tertiary care hospitals.[10,35,37,38]

The medical tourism industry in India traces its roots to Chennai—specifically, the Apollo Hospitals group, a private entity, which opened its first international branch in 2007. Navigating the traffic-choked streets of Chennai, it is hard to miss the handsome, coffee-colored marble building with its gold-emblazoned “Apollo” moniker, towering above the hand-painted shops and stalls. Apollo's success inspired the creation of private hospitals and “super-specialty” centers in Chennai offering everything from liposuction to laparoscopic, minimally invasive joint replacement. There is the Shankara Netralaya Eye Institute; MV Hospital for Diabetes and Diabetes Research Center; Madras Ear Nose & Throat Research Foundation, Heart Institute and Institute of Cardiovascular Diseases, all of which are privately owned and nationally recognized.[39–42]

These hospitals steered India's most lucrative health trends, including telemedicine and medical outsourcing, and have received critical acclaim as the home of various medical “firsts” in Asia and India. Examples include the first successful transmyocardial revascularization laser surgery in 1994, the concept of magnetopexy in 1988 and the first successful heart-lung transplant in 1995, to name a few.[43,44] The effect of modernized medicine is seen even in such seemingly low-tech places such as the Chennai Central railway station, where there is a telemedicine facility complete with EKG machines and virtual consultation stations for those passengers who suddenly feel a bout of chest pain along the main concourse.[45]

Glamour and wealth in Indian health care transpire primarily in the private sector, which generates over 70% of all health care revenue (or

6% of GDP in 2005). Highlighting the regional dominance of health care by the private sector, the *South Asian Economic Report* notes that "the proportion of private health expenditure to total health expenditure in the [South Asian] region surpasses that of most countries in the world."[43,46]Contrast this with the US, where the government's share of per capita health care spending tops 50%.

The private sector's starring role in India's health care stems from the dynamics of supply and demand: Private providers' domain encompasses services from x-rays and MRIs to treatment of childhood diarrhea and malaria to prescription drugs. In contrast, government spending on health care actually *decreased* over the past decade.[10,43] Paucity of funding sustains inefficiency in government hospitals, where shortages of drugs, supplies, and personnel persist. One article describes Chennai's Government General Hospital, which sees 10,000 patients per day, as having a workforce "not even adequate to cater to a fourth of these numbers."[43] Such basic deficiencies prompted an acrid editorial comment in the city's major newspaper, *The Hindu:* "It is an ironic outcome of neo-liberal economic reforms that in spite of fundamental policy failures in public health, India is increasingly seen as an attractive international health care destination."[38]

FIRST ENCOUNTER

The couple walking into the room comprises an elderly man and woman who instinctively seem to lean on each other for support. The woman is slightly heavyset, with salt-and-pepper hair loosely threaded into a braid and secured with jasmine flowers that have begun to brown in the midday heat. She absently pats these flowers as she looks up respectfully at Dr Raman. Her husband, a stocky, dark gentleman wearing a white*dhoti* and a rather ill-fitting button-down shirt, is carrying a bulky briefcase that he sets down before cupping his hands together. *"Vanakkam,* doctor," he says with genuine reverence and enthusiasm. It is only two words—hello, doctor—but the undertone of hope in his voice is almost palpable.

Because there isn't an extra chair, I am trying to stand as inconspicuously as possible under the x-ray viewing box. I wonder what patients will think of my rather random presence as they discuss their aches and pains. Would they request that I leave? Demand an introduction? Should I introduce myself—exposing my broken, American-accented Tamil—or wait for Dr Raman to take the lead?

As I find out, introductions are apparently unnecessary. Patients seem to think that as long as Dr Raman accepts my presence, they don't need to know who I am, or what my qualifications are for listening to their concerns. Indeed, throughout my rotation here only one patient actually looks at me directly and asks, "And so, who exactly are you?" It's somewhat of a relief in a sense; part of me fears that by opening my mouth and talking too much, my status as foreigner will instantly be revealed. Throughout the rotation, I grow used to my place as a nameless physician-in-training, nodding in sympathy during the discussions of pain and suffering, hanging up x-rays on the lightboard, helping frail patients climb down from the table.

Now, Mr Vasu launches into a description of his wife's agonizing back pain. A few sentences in, however, Dr Raman stops him. "She's the one with the pain?" he asks in Tamil. Both nod hesitantly, and Dr Raman motions to the lady. "Then you talk," he says pointedly. Mrs Vasu looks taken aback for a second, reaching up to her jasmine flowers for support. But she begins, after a moment, to describe the dull ache in her lower back that has disturbed her for the past two years, now intensifying to the point where she can no longer sleep. Dr Raman nods in understanding and is already scribbling down something on a prescription pad. "Let's see the x-ray?" he half-asks, half-commands as he signs the pad with a flourish.

Mrs Vasu looks at her husband expectantly; he readily pulls a large manila envelope from his briefcase. Across the front is stamped in faded gray print: "Swaminathan Scans, Ltd." The envelope contains a glossy x-ray of what is apparently Mrs Vasu's spine. As I will discover throughout the month, even the poorest of patients come to clinic with "high-*funda*"[b] radiologic images capturing their anatomy. Obtaining a computed tomography scan or x-ray is as easy as going into a neighborhood stall and paying a few dollars, depending on where the scan is obtained. Meanwhile, if a patient is not able to afford these prices, Dr Raman has agreements with specific x-ray and MRI centers that will perform the tests free of cost.

Dr Raman and I look at Mrs Vasu's x-ray. He asks me what I see; I gesture, with weatherman-style vagueness honed during third-year rotations, over an area with joint-space narrowing. He nods thoughtfully, studies the scan for a few more moments and then switches off the viewing light. "Right," he tells the couple, who seem to be hanging on his every word. "Absolutely nothing to worry about. No TB, no infection. Take these medicines, and I will give you some exercises; it will become all right." He slides over a prescription (for what looks like an NSAID, PPI, and a topical muscle relaxant) as Mr and Mrs Vasu nod vigorously in tandem, relief evident in both their faces.

Dr Raman stands up. "Also, you will need to lose five kilos," he says bluntly. "You're overweight, and your pain will improve if you lose this weight." Mrs Vasu and her husband look genuinely surprised, almost as though the concept of having "too much" weight is foreign. Dr Raman briskly continues as he hands her a card. "My dietician— Jyotsna—she is very good. You can make an appointment with her, and she will give you a diet. Then come and see me day-after-tomorrow. I will show you some yoga exercises to help the pain." She mouths agreement, looking at her husband with a slightly bemused expression.

Side Effects of Modernization: New Challenges to India's Disease Burden

As the lifestyle of material success touches more of India's citizens, so also does the sobering impact of a new set of health conditions. They are a cohort of "lifestyle" diseases: metabolic syndrome, Type II diabetes, coronary artery disease, obesity, and tobacco and alcohol addiction.

In 2004, almost half the disease burden in South Asia consisted of noncommunicable diseases, a 10% increase over the past decade. Over the next decade, these "lifestyle" diseases will comprise an estimated 57% of India's total burden of disease—with heart disease rising as India's number one killer—whereas infectious diseases' share will sink below 24%.[10,47,48]

Heart disease already kills 15% of India's population each year, whereas over a third have metabolic syndrome. The Public Health Foundation of India notes: "India tops the world list in terms of the disability burden due to heart and blood vessel disease (more than all industrial countries put together)."[49] Tobacco use persists as a major killer, responsible for India's world dominance in oral cancer prevalence. And noninsulin-dependent diabetes mellitus (Type II Diabetes) is also a health scourge for Indians: With over 12% of its adult population developing the disease each year, India is the "diabetes capital" of the world, housing more diabetic patients than any other nation.[10,48-50]

The changing disease demographic is rooted partly in lifestyle transitions—urbanization, sedentary behavior, changing food choices—which escalate classic risk factors causing diabetes and heart disease. Where Indian residents once feared undernutrition, they are now caught in a growing epidemic of over-nourishment: obesity, and its anvil of chronic disease. A large cohort study found nearly half of Chennai's urban population was obese by a standard of BMI > 25; adopting a waist-circumference-based definition (waist size greater than 90 cm) placed even more Chennaites—55% of Chennai females—in this category.[51] Another study identified 16% of urban Indian schoolchildren as overweight, with close to a third demonstrating insulin resistance (a precursor to diabetes).[52] Misra et al comment that "the rapid nutritional and lifestyle transition in urbanized areas ... are prime reasons for increasing prevalence of obesity and the metabolic syndrome."[50]

More evidence supports a coupling of urbanization and lifestyle disease. In urban Chennai, high socioeconomic status was found to be an independent, statistically significant predictor of being overweight or obese.[51,52] Lifestyle may also be partly responsible for hypertension in nearly a third of Chennai's urban residents; a Madras Medical Institute cross-sectional study of 2007 Chennai-based volunteers found that higher monthly income correlated positively with blood pressure. In this study, belonging to a middle-class classification or higher increased the chance of having high blood pressure by 150%.[50,52-54]

Hastened by obesity and metabolic syndrome, diabetes is a particularly problematic component of the "lifestyle epidemic" in urban India. A large cohort study from Chennai's MV Diabetes Center (Chennai Urban and Rural Epidemiologic Study, CURES) found that diabetes prevalence in urban Chennai increased by 72.3% from 1989 to 2004.[55] Although this increase arises partly from improved methods of detection and earlier diagnosis, lifestyle factors are also contributors. The Diabetes Prevention Program underscored the role of obesity, lack of physical activity, and poor diet in exacerbating cardiovascular disease and diabetes. Various epidemiologic studies even quantify this effect, suggesting obesity is responsible for up to 90% of the risk of acquiring type 2 diabetes.[56] As the incidence of obesity and impaired glucose tolerance rises in a younger Indian population, the number of Indian diabetic patients inevitably increases.

Urbanization has clearly transformed India's health, but new evidence suggests genes may also contribute to Indians' growing burden of chronic disease. Of note, India has a higher prevalence of diabetes and cardiovascular disease than other Asian industrializing nations. Factors hastening cardiovascular disease—high cholesterol, cell markers of inflammation, obesity and overweight, endothelial dysfunction (disruption in the lining of blood vessels leading to formation of artery-clogging plaques), thrombosis (clots which can block blood flow, leading to heart attacks), glucose intolerance (a precursor to dia-betes)—affect a greater proportion of South Asians than Caucasians, with onset in Indians occurring 10 to 15 years earlier. Nearly half of all cases of heart disease are detected in Indians younger than age 50— and over half of all cardiovascular deaths occur in people <70 years in India, compared with 22% in developed countries.[50,55,57,58]

Several studies extricate the genetic component to lifestyle diseases in India. One study by the International Diabetes Epidemiology Group showed that Indians' risk of acquiring diabetes increased at lower levels of body mass index (BMI) compared with Europeans—and is more sensitive to smaller increases in BMI.[59]The National Urban Indian Survey found central obesity had a higher prevalence (50%) among Indians; even lean-BMI individuals tended to have central obesity and high percentages of body fat. This is particularly significant as central adiposity—the so-called "apple shape"—carries the strongest relationship to impaired glucose tolerance in Indians.[50,55,57,58,60,61] In fact, the "unique" Indian overweight and obesity patterns prompted the Association of Physicians of India to issue revised 2009 guidelines for obesity and metabolic syndrome, aimed at identifying more at-risk Indians and staving off disease through early prevention strategies.[62] In sum, such research supports a theory that Indians may have inherent genetic susceptibility to diabetes and cardiovascular disease, now unmasked by lifestyle changes accompanying India's urbanization and industrialization.

Co-pay

As the Vasus stand to leave, Mr Vasu asks, *"Evlovu,* Doctor?" [How much, Doctor?] Dr Raman waves them off. "You can pay me later; it will be 500 rupees." Mr Vasu cups his hand once more in respect, and the couple exits the room. I am struck by this trust-based system of accounting—when exactly is "later"?—as I begin straightening the room for our next patient.

Paying for India's Modernizing Health Care

Most citizens pay out of pocket for health care (40% of the $27 per capita health spending in 2002). Talking to various patients during my rotation, I gleaned that a typical middle-class Chennai household (mean income INR 10,000) generally finds outpatient medical care at a government hospital to be affordable: A clinic visit costs anywhere from $.50 to $5; a knee x-ray can be obtained for $4; a week's supply of antibiotics will run less than $3.[63]

Unfortunately, current trends in health and social demographics are already driving up prices, threatening to render an era of affordable health care obsolete. First, as private tertiary care hospitals flourish in the wake of the medical tourism goldmine, state-of-the-art procedures and facilities demand increasingly prohibitive prices. Second, as the prevalence of "lifestyle" diseases increase, so also does the cost of receiving treatment. In contrast to times when disease treatment entailed an empiric course of relatively cheap chloramphenicol, today's urban diseases demand expensive antidotes—cardiac catheterizations and stents, a multidrug regimen of blood pressure- and cholesterol-lowering medications, coronary bypass procedures, and total knee replacements.

Thus, staying healthy is getting expensive in urban India. A five-year cohort study of 556 diabetic patients found that total household expenditures on health care—constituting 34% of median annual income—increased 113% during a five-year period, with the *highest* increase in percentage outlay occurring in the lowest socioeconomic group.[64] Another study found that 50% of patients hospitalized in 2004 faced hospital expenditures exceeding 10% of annual household spending.[64] Reaching deeper into their pockets to receive medical care, India's citizens often must sell their assets or borrow to meet costs—as was the case for 40% of Tamil Nadu's inpatients in 2007.[10]

Debriefing

When Dr Raman and I sit down later to debrief, I ask him if he thinks his patients will take the medications he prescribes, make the appointments he recommends or follow-up on his instructions. "Oh, they do," he tells me matter-of-factly. "It is a different doctor-patient relationship. Patients here take these appointments very seriously. It is almost never the case that a patient will not schedule a follow-up appointment or fail to obtain the scans needed or ignore the prescriptions. They want to get better, of course, and they believe that will come from listening to the doctor."

I think back to my small repertoire of patients encounters as a medical student in Michigan: the hour spent answering the questions of a pneumonia-ridden businessman who is concerned about taking a new antibiotic, explaining to an irate sickle-cell patient the rationale for capping her morphine dosage, reiterating to a frustrated patient that her work-up to date has not yet revealed a specific etiology to "fix." As I will find throughout the month, it *is* a different world here. Dr Raman does very little explaining; the patients, in turn, question and challenge very little. They accept the added medications, the reassurance that "it will become all right," the exhortation to "get an MRI scan and follow-up in a week," almost as if it is a duty. The inherent trust they place in Dr Raman's opinion is, perhaps, reflective of the prevailing cultural attitudes toward physicians. As a patient stated: "For the majority of us, a doctor is virtually God—one who is beyond questions or doubts and has solutions to all our ills."

Reflections: The Future of Indian Health Care

The opulent luxury of private-sector hospitals, juxtaposed with creeping improvement in basic health indicators, hints at the dichotomy of health care in India. Even as efforts abound to quash pathogens and parasites—the vestiges of underdevelopment and poverty—India's private health care industry flourishes, bringing with it the promise of profits. This twin agenda operates at polar opposites of socioeconomic class, two seemingly disparate foci running in parallel. Is it sustainable—and equitable? The Asian Development Bank's South Asian Economic Report warns, "Although medical outsourcing will give impetus to economic growth in the region, it could also distort the availability of medical care away from South Asia's poor as the health systems cater to clients from the developed world."[10] Growing evidence supports the troubling emergence of a two-tier system, whereby quality health care caters to and becomes the de facto privilege of the upper class, and the average citizen depends on underfunded, understaffed public facilities. This, in turn, portends a spiral of suboptimal health for the nation's poor and middle class, carrying somber ramifications for goals of public welfare and social equality.

The dramatic changes in India's health environment and shift of the disease profile presage an economic and social transformation in health care delivery. On the one hand, the nation's newfound riches promote certain types of health, effacing the disease-ridden India stereotypes of middle-school geography books and quaint Rudyard Kipling tales—where epidemics of cholera and polio consumed millions and curable infectious diseases terrorized the lives of city inhabitants. But even as strides are made in the realm of hygiene and hospital infrastructure, new health challenges emerge in the shifting face of disease, cost containment, and health care access. Such challenges carry relevance to health care in the developing world, as they represent the prototypical public health needs of a nation straddling the realities of persistent poverty and the heady success of breakneck growth.

Reflections: The Beginning of a Journey

I am waiting for takeoff. It is pitch black outside—approximately 1:30 am, Indian Standard Time—and my head is heavy with fatigue. The dry Lufthansa cabin air keeps forcing me to sneeze, thwarting halfhearted attempts at sleep. My mind also appears to be part of the plot to keep me awake, buzzing with a flurry of thoughts and impressions of a summer in India.

Here lies a crucial area for future medical service in Chennai: public health education, enabling its citizens to understand—and practice—the elements of a fundamentally long and healthy life.

I revisit the trepidation that pervaded my first bumpy car ride to Dr Raman's clinic. I had imagined— and feared—the rapidly urbanizing, technology-championing, café-laden “new” Chennai would serve as a rude shock, alien from the “Madras” of childhood visits. I had wondered if any trace of my heritage—the culture that sparked my passion for art, music, and dance—might remain in this revamped, modernizing pantheon of software outsourcing. And, with quavering hope and resolve, I had entered this city with the goal of serving in a medical capacity, of understanding the city's unique health needs.

Chennai had changed. There were glitzy new stores, air-conditioned restaurants, new highways, the flood of bright matchbox Fords and Hondas. There existed a growing sense of empowerment: young professional women confidently rode to work on scooters and mopeds in the midst of rush-hour traffic; billboards heralded the grand opening of new technology parks; newly installed garbage cans on street corners exhorted in Tamil: “Don't Litter: Keep Chennai Beautiful.” And, of course, there were momentous changes in health care. Multistorey hospitals towered over dilapidated clinics; clean public restroom facilities emerged as reliable fixtures in malls and restaurants. Complex surgeries no longer necessitated expensive trips abroad; instead, medical tourism now brought thousands of foreigners and a steady stream of profits to Chennai's hospitals each year.

And yet, in Chennai much remains the same. The ancient temples I had visited on trips past are still as ancient as ever. The homemade *palgova* and cardamom milk from my favorite (non-air-conditioned) dairy store still tastes as divine as I remembered it, and unfortunately is still as fattening. Cows still lazily ruminate as they always have on pot-holed side roads, unperturbed as surrounding cars unleash a blaring cacophony of honks. The unique aroma of incense, spices, humidity, car exhaust, roasted peanuts, and coffee powder still hangs in the air—it's simply mixed with more exhaust fumes.

Chennai's health also sadly carries echoes of the past. Beggars still cry for food on street corners; slums still spill human filth and suffering. Public health initiatives have mitigated—but not eliminated—once epidemic communicable diseases. India still carries the world's greatest burden of patients with tuberculosis, and must face the challenge of emerging multidrug-resistant strains. Polio and measles may no longer consume lives in Chennai, but HIV/AIDS and malaria persist as important health concerns; infectious diarrhea still kills an unacceptable number of children each year. The battle to provide fundamental public health needs is still not won, and it must not be ignored in favor of the *haute* trend of elite health care.

India's paradox of constancy and change—epitomized by its health care—serves as Chennai's major theme. As a student in Dr Raman's clinic, I interacted with a truly diverse socioeconomic, cultural, and religious cross-section of Chennai society, many of whom were crippled by pain exacerbated by a sedentary lifestyle, excess weight, and poor dietary habits. Often, patients had no idea that they were overweight—or that lifestyle could potentially undermine their health. Several patients were unfamiliar with the concept of a “heart-healthy” diet, or unaware that they could modify their risk for diabetes and heart attacks with simple dietary and activity modifications. Here lies a crucial area for future medical service in Chennai: public health education, enabling its citizens to understand—and practice—the elements of a fundamentally long and healthy life.

The plane revs up its engines in preparation for the final rapid acceleration before takeoff. As we rise smoothly above the twinkling lights of Chennai, I peer out the window, straining for a final view of the city and its roads (which are, even at this hour, packed with cars), exhaling a wave of nostalgia as the city lights fade, covered by growing patches of clouds and mist. I press my face against the cool window, imagining I can still see a twinkling light or two.

As the last visible light disappears from view, I realize it is only a matter of time before I return. For the past month has introduced me to a challenge—a medical need that speaks to my interests in public health, service, and culture—that I am determined to revisit and tackle as a physician.

Even with all its chaos and inefficiency, Chennai somehow, miraculously, inexplicably, stubbornly, still works. It's reassuring, for I know it will still be working when I fly back at some point in the future into the humid air surrounding Anna International Airport.

I wave goodbye one last time before pulling the window cover shut. *"Paarkahm,"* I whisper, to no-one in particular. It's a promise: See you soon.

FOOTNOTES

[a] *Pradakshinam:* walking around the temple in a sign of respect—40 *pradakshinams,* depending on the size of the temple, can equal several miles!

[b] *High-funda:* a slang term used to capture the idea of "high-tech," "new-age," or "chic"—as in: "That's one high-funda Lexus IS Coupe you're driving!"

DISCLOSURE STATEMENT

The author(s) have no conflicts of interest to disclose.

DREAMS OF LIVING MEN

If there is one place on the face of earth where all the dreams of living men have found a home from the very earliest days when man began the dream of existence, it is India.

— Romain Rolland, 1866-1944, French writer and art historian, winner of the 1915 Nobel Prize for Literature

REFERENCES

1. District profile: Census 2001 Data [table on the Internet]. Chennai, India: Tamil Nadu Government. 2001. [cited 2010 Feb 10]. Available from: www.chennai.tn.nic.in/chndistprof.htm#CENSUS.

2. Census of India: Housing profile: Tamil Nadu (33) [tables on the Internet]. New Delhi, India: Office of the Registrar General and Census Commissioner. 2001. [cited 2010 Feb 10]. Available from:www.censusindia.gov.in/Census_Data_2001/States_at_glance/State_Links/33_tn.pdf.

3. tn.gov.in. [home page on the Internet] Chennai, India: Government of Tamil Nadu; [updated 2010; cited 2010 Feb 10]. Available from: www.tn.gov.in.

4. Los Angeles (city) California. [table on the Internet.] US Census Bureau: State and County QuickFacts, Data derived from Population Estimates. 2000. Census of Population and Housing; 2009 Nov [cited 2010 Feb 14]. Available from: http://quickfacts.census.gov/qfd/states/06/0644000.html.

5. December Music Festival [home page on the Internet]. Chennai, India: The Music Academy of Chennai.2009. [cited 2010 Feb 10]. Available from: www.musicacademymadras.in/december_music_festival.php.

6. Chennai most attractive city for offshoring services [monograph on the Internet]. Chennai, India: The Hindu Business Line. 2006. Apr 2 [cited 2010 Feb 10]. Available from: www.thehindu-

businessline.com/2006/04/02/stories/2006040202550100.htm.

7. Ablett J, Baijal A, Beinhocker E, et al. The 'Bird of Gold': The Rise of India's Consumer Market [monograph on the Internet]. San Francisco, CA: McKinsey Global Institute; 2007 May. [cited 2010 Feb 10]. Available from:www.mckinsey.com/mgi/reports/pdfs/india_consumer_market/MGI_india_consumer_full_report.pdf.

8. Raja M. Chennai moves into the big leagues [monograph on the Internet]. Hong Kong: Asia Times Online; 2007 Mar 8. [cited 2010 Feb 10]. Available from:www.atimes.com/atimes/South_Asia/IC08Df04.html.

9. Indian States: Economy and Business—Tamil Nadu [monograph on the Internet]. Gurgaon (Haryana), India: India Brand Equity Foundation: Confederation of India Industry. 2005. [cited 2010 Feb 10]. Available from: www.ibef.org/download%5CTamilNadu_may06.pdf.

10. South Asia economic report [monograph on the Internet]. Mandaluyong City, Phillipines: Asian Development Bank. 2007. Jun 9 [cited 2010 Feb 10]. Available from:www.adb.org/Documents/Reports/Social-Sectors-Transition.pdf.

11. 5 cities improve their ranking: survey [monograph on the Internet]. New Delhi, India: The Economic Times. 2007. Mar 14 March [cited 2010 Feb 10]. Available from: http://economic-times.indiatimes.com/5_cities_im-prove_rankings_in_living_standards/articleshow/1764200.cms.

12. Mercer's 2009 quality of living survey highlights—global [monograph on the Internet]. New York: Mercer Human Resources Consulting. 2009. Apr 28 [cited 2010 Feb 10]. Available from:www.mercer.com/qualityofliving.

13. Statistical hand book 2010 [monograph on the Internet] Chennai, India: Government of Tamil Nadu: Department of Economics and Statistics. 2010. [cited 2010 Feb 10]. Available from: www.tn.gov.in/deptst/.

14. Part E: Select socioeconomic indicators for all states and India [tables on the Internet. Chennai, India: Government of Tamil Nadu: Department of Economics and Statistics; 2010 [cited 2010 Feb 10]. Available from: www.tn.gov.in/deptst/economicindi-catorall.pdf.

15. D'Souza V. Our community, our responsibility. Bangalore, India: India Together Newsletter Online [serial on the Internet] 1999 Jan. [cited 2010 Feb 10]. Available from: www.indiatogether.org/stories/beach.htm.

16. Tamil Nadu Health Systems Project [monograph on the Internet]. Chennai, India: Government of Tamil Nadu: Department of Health and Family Welfare. 2005-2009. [cited 2010 Feb 10]. Available from:www.tnhsp.org/.

17. tnmsc.com. [home page on the Internet]. Chennai, India: Tamil Nadu Medical Services Corporation. © 2009. [cited 2010 Feb 10]. Available from: www.tnmsc.com/tnmsc/new/index.php.

18. Pradesh A. Telegu Ganga water released for Chennai. 2007 Aug 9. Chennai, India: The Hindu Online [serial on the Internet]. [cited 2010 Feb 10]. Available from:www.hindu.com/2007/08/09/stories/20070809548004 00.htm.

19. Department of Public Health and Preventive Medicine. [monograph on the Internet.] Health and Family Welfare Department, Government of Tamil Nadu. Table #2 on vaccine preventable diseases. 2003-2009. [cited 2010 Feb 14]. Available from: www.tnhealth.org/dphis.htm.

20. Disease burden: Malaria incidence in rural and urban areas of Tamil Nadu [table on the Internet] In: Malaria: Frequently asked questions. Chennai, India: Health and Family Welfare Department: Government of Tamil Nadu; © 2009 [cited 2010 Feb 10]. Available from:www.tnhealth.org/dphfacts/malaria.htm#diseaseburden.

21. Current Leprosy Situation as on April 2008 [monograph on the Internet]. New Delhi, India: National Leprosy Eradication Programme. 2008. Apr [cited 2010 Feb 10]. Available from: http://nlep.nic.in/data-2008.htm.

22. Table 14.5 Services rendered by the medical institutions, 2008-2009 [table on the Internet.] Chennai, India: Government of Tamil Nadu, Department of Economics and Statistics: Statistical Hand Book; 2010 [cited 2010 Feb 10]. Available from: www.tn.gov.in/deptst/Stat.htm, click on 14 Medical and Health.

23. Table 14.6. Population, beds and doctors ratio, districtwise 2008-2009 [table on the Internet]. Chennai, India: Government of Tamil Nadu, Department of Economics and Statistics: Statistical Hand Book; 2010 [cited 2010 Feb 10]. Available from: www.tn.gov.in/deptst/Stat.htm, click on 14 Medical and Health.

24. Profile of Tamil Nadu: Annual report [monograph on the Internet] New Delhi, India: Ministry of Health and Family Welfare; 2008 Mar. Available from: http://mohfw.nic.in/NRHM/State%20Files/tamilnadu.htm.

25. Chennai district - Statistical Handbook. 2006: 45. [tables on the Internet]. Chennai, India: Chennai District. 2006. Available from: http://chennai.gov.in/shb-main.htm.

26. Tamil Nadu. Data Highlights: The Scheduled Castes. Census of India 2001 [monograph on the Internet]. New Delhi, India: Office of the Registrar General and Census Commissioner. 2001. [cited 2010 Feb 10]. Available from: http://censusindia.gov.in/Tables_Published/SCST/dh_sc_ tamilnadu.pdf.

27. Valar Kalvi scheme has helped state increase literacy rate. 2009 Feb 24. Chennai, India: The Hindu Online [newspaper on the Internet]. [cited 2010 Feb 10]. Available from:www.thehindu.com/2009/02/24/stories/2009022459600400.htm.

28. Srinivasan M. Toilets, yes; but usable? 2007 Jul 3. Hardly. Chennai, India: The Hindu Online [newspaper on the Internet]. [cited 2010 Feb 10]. Available from:www.hindu.com/2007/07/03/stories/2007070358220300.htm.

29. Tamil Nadu and Kerala [monograph on the Internet]. New York: UNICEF. 2009 Aug 5. [cited 2010 Feb 10]. Available from: www.unicef.org/india/state_profiles_4352.htm.

30. Krishnakumar A. A Sanitation Emergency. Frontline Magazine Online [serial on the Internet] 2003 Nov 22 - Dec 5. [cited 2010 Feb 10];20(24):[about 3 p]. Available from:www.hinduonnet.com/fline/fl2024/stories/20031205002510100.htm.

31. Bhandari M. Trapped in Stench. Connect! [serial on the Internet] 2004 Dec 6. [cited 2010 Feb 10]: [about 3 p]. Available from: www.sawf.org/newedit/edit12062004/infocus.asp.

32. Chandramouli C. Slums In Chennai: A Profile [monograph on the Internet]. Ontario, Canada: York University Department of Environmental Studies; 2003 Dec 15-17. [cited 2010 Feb 10]. Available from:www.yorku.ca/bunchmj/ICEH/proceedings/Chandramouli_C_ICEH_ papers_82to88.pdf.

33. Tamil Nadu—An Economic Appraisal 2002-2003. Department of Evaluation and Applied Research, Chapter 12. Publication available online at www.tn.gov.in/dear and section 12 atwww.tn.gov.in/dear/Urban_devpt.pdf.

34. krishnaraman.com. [home page on the Internet] Chennai, India: www.krishnaraman.com; © 2008 [cited 2010 Feb 10]. Available from: www.krishnaraman.com/profile.php.

35. Pandeya R. Healthcare goes five-star in India. 2006 Dec 2. Rediff India Abroad Magazine Online [serial on the Internet] [cited 2010 Feb 10; about 4 p]. Available from:www.rediff.com/money/2006/dec/02spec2.htm.

36. indiamedicaltourism.net [homepage on the Internet]. India Medical Tourism India © 2005 [cited 2010 Feb 10]. Available from: www.india-medicaltourism.net/medical_tourism_india_news/index.html.

37. Srivastava N. The road ahead [monograph on the Internet]. Mumbai, India: Express Healthcare Management; 2002 Nov 16-30. [cited 2010 Feb 10]. Available from:www.expresshealthcaremgmt.com/20021130/focus2.shtml.

38. Ananthakrishnan G. Boom time for Medicare. The Hindu Magazine online [serial on the Internet]; 2006 Apr 30. [cited

2007 Aug 15];[about 3 p]. Available from:www.hindu.com/thehindu/mag/2006/04/30/stories/2006043000010100.htm.

39. apollohospitals.com. [home page on the Internet] Chennai, India: Apollo Hospitals; [cited 2010 Feb 10]. Available from: www.apollohospitals.com.

40. sankaranethralaya.org; [home page on the Internet]. Chennai, India: Sankara Nethralaya: updated 2010 Feb 12 [cited 2010 Feb 15]. Available from: www.sankaranethralaya.org/.

41. mvdiabetes.com. [home page on the Internet]. Chennai, India: M V Hospital for Diabetes (P), Ltd; 2007 [cited 2010 Feb 15]. Available from: www.mvdiabetes.com/

42. srmc.edu/medicalcentre.htm. [home page on the Internet]. Chennai, India: Sri Ramachandra Medical Centre; 2010 [cited 2010 Feb 10]. Available from: www.srmc.edu/medicalcentre.htm.

43. Krishnakumar A. A Gateway to Health. Frontline Magazine Online [serial on the Internet] 2003 May 24 - Jun 6. [cited 2010 Feb 10];20(11):[about 3 p]. Available from: www.hinduonnet.com/fline/fl2011/stories/20030606003811000.htm.

44. Kannan R. Chennai hospitals unveil 'pinhole' surgery. The Hindu [serial on the Internet] 2007 May 12. [cited 2010 Feb 10]. Available from: www.thehindu.com/2007/05/12/stories/2007051205110500.htm.

45. Telemedicine. [monograph on the Internet]. Chennai, India: Sri Ramachandra Medical Centre; 2010 [cited 2010 Feb 10]. Available from: www.srmc.edu/telemedicine.htm.

46. Heitzman J, Worden RL. Health Care: Role of the Government. 1995. In: India: A Country Study. Washington, DC: GPO for the Library of Congress.

47. Ramachandran A, Snehalatha C, Mary S, Mukesh B, Bhaskar AD, Vijay V. Indian Diabetes Prevention Programme (IDPP). The Indian Diabetes Prevention Programme shows that lifestyle modification and metformin prevent type 2 diabetes in Asian

Indian subjects with impaired glucose tolerance (IDPP-1)Diabetologia 2005 Jan. 4;49:289–97.

48. Misra A, Khurana I. Obesity and the metabolic syndrome in developing countries. J Clin Endocrinol Metab. 2008 Nov;93(11 Suppl 1):S9–30.

49. Why is being physically active important for urban Indians?. [monograph on the Internet]. New Delhi, India: Public Health Foundation of India; 2008-2009 [cited 2010 Feb 10]. Available from: http://healthy-india. org/beingactive1.asp.

50. Misra A, Misra R, Wijesuriya M, Banerjee D. The metabolic syndrome in South Asians: continuing escalation & possible solutions. Indian J Med Res. 2007 Mar;125(3):345–54.

51. Deepa M, Farooq S, Deepa R, Manjula D, Mohan V. Prevalence and significance of generalized and central body obesity in an urban Asian Indian population in Chennai, India (CURES: 47) Eur J Clin Nutr.2009 Feb;63(2):259–67.

52. Mohan V, Shanthirani S, Deepa R, et al. Intra-urban differences in the prevalence of the metabolic syndrome in southern India—the Chennai Urban Population Study (CUPS No. 4) Diabet Med. 2001 Apr;18(4):280–7.

53. Subramaniam SV, Smith GD. Patterns, distribution, and determinants of under- and overnutrition: a population-based study of women in India. Am J Clin Nutr. 2006 Sep;84(3):633–40.

54. Chockalingam A, Ganesan N, Venkatesan S, et al. Patterns and predictors of prehypertension among “healthy” urban adults in India. Angiology. 2005 Sep-Oct;56(5):557–63.

55. Mohan V, Deepa M, Deepa R, et al. Secular trends in the prevalence of diabetes and impaired glucose tolerance in urban South India—the Chennai Urban Rural Epidemiology Study (CURES-17) Diabetologia.2006 Jun;49(6):1175–8.

56. Hossain P, Kawar B, El Nahas M. Obesity and diabetes in the developing world—a growing challenge.N Engl J Med. 2007 Jan 18;356(3):213–5.

57. Ramachandran A. Epidemiology of diabetes in India—three decades of research. J Assoc Physicians India. 2005 Jan;53:34–8.

58. Misra A, Khurana L. The metabolic syndrome in South Asians: epidemiology, determinants, and prevention. Metab Syndr Relat Disord. 2009 Dec;7(6):497–514.

59. Hitman GA, McCarthy MI, Mohan V, Viswanathan M. The genetics of non-insulin-dependent diabetes mellitus in south India: an overview. Ann Med. 1992 Dec;24(6):491–7.

60. Ramachandran A, Snehalatha C, Kapur A, et al. Diabetes Epidemiology Study Group in India (DESI). High prevalence of diabetes and impaired glucose tolerance in India: National Urban Diabetes Survey.Diabetologia. 2001 Sep;44(9):1091–101.

61. Snehalatha C, Sivasankari S, Satyavani K, Vijay V, Ramachandran A. Insulin resistance alone does not explain the clustering of cardiovascular risk factors in southern India. Diabet Med. 2000 Feb;17(2):152–7.

62. Misra A, Chowbey P, Makkar BM, et al. Concensus Group. Consensus statement for diagnosis of obesity, abdominal obesity and the metabolic syndrome for Asian Indians and recommendations for physical activity, medical and surgical management. J Assoc Physicians India. 2009 Feb;57:163–70.

63. Services [monograph on the Internet]. Chennai, India: Tamil Nadu Medical Services Corporation. 2009. [cited 2010 Feb 10]. Available from: www.tnmsc.com/tnmsc/new/html/services.php#.

64. Ramachandran A, Ramachandran S, Snehalatha C, et al. Increasing expenditure on health care incurred by diabetic subjects in a developing country. Diabetes Care. 2007 Feb;30(2):252–6.

12
CHAPTER

Key Considerations for an Economic and Legal Framework Facilitating Medical Travel

Saba Hinrichs-Krapels,[1,*] Sarah Bussmann,[2] Christopher Dobyns,[3] Ondřej Kácha,[4] Nora Ratzmann,[5] Julie Holm Thorvaldsen,[6] and Kai Ruggeri[7,8]

[1] *The Policy Institute, King's College London, London, UK,*

[2] *Department of Psychology, University of Cambridge, Cambridge, UK,*

[3] *Department of Politics and International Relations, University of Oxford, Oxford, UK,*

[4] *Department of Psychology, Faculty of Arts, Masaryk University, Brno, Czech Republic,*

[5] *Department of Social Policy, London School of Economics, London, UK,*

[6] *Department of Psychology, Norwegian University of Science and Technology, Trondheim, Norway,*

[7] *Policy Research Group, Department of Psychology, University of Cambridge, Cambridge, UK,*

[8] *Engineering Design Centre, Department of Engineering, University of Cambridge, Cambridge, UK*

ABSTRACT

Introduction

In contrast to medical tourism, which broadly describes the movement of individuals for non-urgent treatments (e.g., cosmetic surgeries) or recreational and leisure purposes (1), the definition of medical travel explicitly refers to patients crossing national borders with the purpose of receiving treatment that has been determined as essential to maintaining quality of life by a health professional, but may not need to be performed urgently (2). The work reported in this paper is part of a broad, comprehensive project aimed at developing a global health access policy (GHAP) to facilitate safety, efficiency, and equal access in medical travel, as outlined by the Organisation for Economic Co-operation and Development (OECD) in the context of medical tourism (1). This work establishes the economic and legal perspective of a GHAP, identifying the key issues to consider when setting up such a policy to ensure economic viability and sustainability, as well as limiting risk to the stakeholders involved.

An increase in medical travel might result in a broad spectrum of consequences, e.g., (a) a wide-ranging impact on health-care systems (3, 4), (b) access of inbound countries (countries which receive medical travelers) to a larger market and additional revenue streams (5), and (c) easier access of patients in outbound countries (countries which send medical travelers) to medical care, reduced costs, and better standard of care (1). However, the uncontrolled movement of patients carries significant risks for both patients and health-care systems with respect to quality standards, dissemination of diseases, financial implications of adverse events, and unregulated developments on the health-care market (6, 7). Accordingly, the costs and benefits of medical travel as well as associated ethical and legal concerns call for a systematic evaluation to ensure the economic viability and sustainability of medical travel.

Several agreements that focus on facilitating medical travel already exist in certain regions, including those between the EU member states (8), within North America (NAFTA), and between OECD members such as the United States and Korea (1). For example, the EU's Patients' Rights Directive supports policies within the EU that ensures that patients making use of medical services abroad are entitled to receive financial reimbursements and follow-up care in their home country (8). However, this is still a directive and not regulation, and therefore, the extent to which financial support is given can differ for each member state. Furthermore, the existing frameworks often fail to consider the much wider needs of patients traveling for care, such as protection and indemnification, psychosocial support during rehabilitation and long-term liability, as well as ensuring access and equity (7, 9). Globally, therefore, medical travel remains a widely unregulated extension of health care (10).

As such, GHAP seeks to address existing gaps in the regulation of medical travel. There is a need for more evidence-based, patient-centered, and population-centered frameworks ensuring positive outcomes for all stakeholders, accounting for potential areas of increased risk and improving sustainability and equity. In the following section, we focus on the economic and legal considerations required for the development of such frameworks.

Consequences and Concerns of Medical Travel

This paper focuses on the economic and legal consequences of medical travel (specified earlier as essential and non-cosmetic care), from the moment a patient starts to consider the option of medical travel, through receiving treatment and posttreatment care, to returning to the outbound country. Our aim was twofold: first, to broadly identify from existing literature the economic and legal risks of medical travel, given its aforementioned global increase, and second, to synthesize these risks and provide key considerations on how to establish a legal and economic framework for medical travel. We note that we did not intend to provide a full breakdown of the costs and benefits of medical travel as this was out of scope of our project and not justified by our broad literature search. A non-systematic review of scientific and gray literature was conducted. Articles were considered if they were found in any of the six databases under the terms "medical tourism" or "medical travel" since the year 2000, with no language restrictions. The databases used were PubMed, EconLit, Google Scholar, the World Bank research database, Europe PubMed Central, and EMBASE.

We first identified primary themes to be reviewed through common arguments used in high-quality sources as well as through common topics in low-quality or potentially very biased material. Papers were included if they provided specific data, policy or practice analyses, or substantial insight to key themes, regarding travel for necessary care. This approach was used to identify primary themes related to economic or legal considerations presented in available literature. Themes identified by the authors are part of a larger study examining medical travel in which various themes relating to medical travel were identified (2). Despite the existence of themes relating to economic and legal frameworks found in the literature, overall, we found that there had been no approach to synthesize this information to specifically inform a GHAP. We have, therefore, summarized the key considerations for establishing an economic and legal framework for GHAP, based on the themes identified. While taking into account that this is a non-exhaustive compilation of proposals, the current work provides the first step toward the development of an economic and legal framework for medical travel (such as GHAP).

Economic Opportunities and Considerations

Medical travel may provide numerous financial incentives for health-care systems and travelers alike. In inbound countries, attracting medical travelers can serve as a mechanism to generate additional revenue (11). The resulting higher utilization of resources, such as hospitals and technical equipment, can contribute to cost reduction in health-care delivery (12). Furthermore, medical travel has the potential to contribute to economic growth through the creation of local job opportunities (13). A study in India investigated the effects of medical travel on the local population and highlighted benefits with regards to enhanced economic growth, revserse brain drain, and the increased investment in medical technologies and training (14). To further maximize the benefits for the local population, tax revenues accrued through linkages with tourism, insurance, food, and hotel business (12) could be allocated to the hospitals and health-care systems for domestic patients (15). Singapore and Cuba are useful examples, since the involvement of the respective countries in medical tourism, according to local authorities, has generated income benefiting the local population (16, 17). There is also evidence to suggest that due to the requirement to comply with international quality standards of care, there has been a trickle-down effect in inbound countries in terms of skills, technologies, and improved quality standards (1, 14). Finally, a liberalization of the market can exert competitive pressure on local health-care providers and consequently drive down costs by encouraging economies to maximize their comparative advantage (5).

Medical travel also has the capacity for negative financial consequences. A significant source of financial concern results from complications associated with medical treatment abroad (7). The potential spread of disease due to the unregulated movement of patients can create significant burden on national health systems and insurance companies (6). Another source of concern is that resources are primarily invested into private hospitals addressing the need of paying medical travelers, resulting in an inequitable two-tiered health system (12, 15, 18). For the local population, this can contribute to rising prices and limited access to health services (12, 15). This phenomenon is seen in Thailand, where a significant number of locals are unable to afford medical care in the private sector as a consequence of artificially high prices targeted toward medical travelers (19). Moreover, it may lead to understaffing of

Table 1. Economic opportunities and risks for medical travel and suggested framework elements.

Inbound countries	
Opportunities	**Suggested framework elements**
1. Generate additional revenue for domestic health-care system	Tax policies to subsidize local health system
2. Create job opportunities	Flexible approach to pricing in inbound countries
3. Tax revenue sharing with domestic health-care system	Accrue tax revenues *via* hotel, food, etc.
Risks	**Suggested framework elements**
1. Spread of disease due to unregulated movement of patients	Improved quality standards
2. The development of an inequitable, two-tiered health system	See inbound opportunities 1 and 3
Outbound countries	
Opportunities	**Suggested framework elements**
1. Residents can seek out treatment that is costly/unavailable	Involvement/engagement in a GHAP
2. Increased availability of specialists	Facilitating more efficient division of labor and knowledge
Risks	**Suggested framework elements**
1. Decreased revenue as a result of patients traveling abroad	See outbound opportunities 1 and 2

the local sector and increased strain on the public-care providers (14). Furthermore, increase in number of high-income patients traveling abroad for care may result in a decrease in revenue in the outbound country, as well as decline in political support for investment and development in local facilities and technology (20). Outbound countries may also have some obligation to subsidize medical travelers from their country, as outlined in the EU's Patients' Rights Directive (8).

Key Elements for an Economic Framework

Based on the factors outlined in the previous section, Table Table11 summarizes the economic risks, opportunities, and suggested key elements in the medical travel context. First, inbound countries must focus on limiting the rise of two-tiered systems. For example, a tax policy could be introduced to health-care providers, which would subsidize the local health-care system. Another solution could entail a flexible approach to pricing of medical procedures. Such interventions would support the local health infrastructure and limit the impact of market mechanisms, which otherwise poses disadvantages to local populations due to the increasing prices (21). Furthermore, establishing a solid economic framework could stimulate competition within inbound countries, which might subsequently result to lower domestic prices. In outbound countries, ensuring that medical travelers are reimbursed for the cost of treatment may increase access to medical care for patients (Verra et al., in review), as well as facilitate follow-up care in the traveler's home country.

Legal Opportunities and Considerations

In order to protect patients and reduce the financial burden on health systems, regulatory oversight needs to be established for each step within the medical travel process. Regarding the early stages of the medical travel process (exploration and planning phases), legal and ethical considerations include providing information for patients, which now includes the booming market of third-party agencies, mediating between patients and prospective medical centers. Such organizations vary in terms of comprehensiveness of service but generally finalize travel arrangements, liaise with the clinics abroad, and make follow-up arrangements (22). As they often function as the first point of contact for patients, they play a significant role as information provider. Currently, there are no regulations in place to monitor the type and reliability of information provided by the third-party agencies, which has led to a market operating in a "regulatory vacuum" (23).

Several studies report shortcomings regarding the communication of medical information and records in the treatment and follow-up stages of travel (24, 25), as well as substandard follow-up care and social support (26, 27). Although the Patients' Rights Directive, in theory, requires that all EU member states provide follow-up care (8), research shows that, in practice, physicians are frequently unaware of this

Table 2. Legal opportunities and risks for medical travel and suggested framework elements.

Inbound countries	
Opportunities	**Suggested framework elements**
1. Adherence to international quality standards	Involvement/engagement in a GHAP
2. Provision of social support during treatment process	
Risks	**Suggested framework elements**
1. Incorrect communication of medical information	Guidelines regarding handling of patient data
2. Discrepancies in international quality standards	Establishment of an authority to oversee standards process
Outbound countries	
Opportunities	**Suggested framework elements**
1. Standardization of services provided by third-party agencies	Involvement/engagement in a GHAP
2. Continuity of care	Implementation of guidelines that ensure follow-up care
Risks	**Suggested framework elements**
1. Financial implications of adverse events and unregulated developments	Outlined responsibilities for each step of medical travel

requirement (28), and additional binding requirements to ensure continuity of care are not in place (9). There is also no consensus on international quality standards, which aim to reduce the differences in the quality level of treatment and hospital facilities (15, 20, 29, 30). As a consequence, there is a greater risk for the transmission of disease as well as the risk of complications pre- and postmedical treatment (9).

Key Elements for a Legal Framework

With medical travelers moving across the world, it is imperative to establish a legal framework clearly outlining responsibilities and liabilities for each step of the medical travel process, especially with regards to transitions such as follow-up care and adverse events. Table Table22 summarizes the legal risks, opportunities, and suggested key elements in the medical travel context. Given the broad nature of these considerations and the inclusion of patient safety and well-being considerations, we note that these include some ethical elements but have given them the general term "legal" considerations. Of particular importance throughout the entire medical travel process should be the definition and implementation of internationally standardized minimum quality standards (30). This process should encompass an external evaluation and associated accreditation program to ensure that health-care facilities comply with minimum standards (23). In order to avoid the emergence of various accrediting bodies and discrepancies in the accreditation process, the establishment of an international authority would be necessary to oversee the process (for instance, the International Society for Quality in Healthcare) (31). Due to the rapidly developing market of third-party agencies, there is an urgent call for an establishment and maintenance of quality standards in the domain of third-party health-care providers (30).

In order to facilitate the treatment process and prevent complications, the communication of medical information is a key element to consider in the development of a legal framework. A failure to transmit medical notes to the treating center abroad could be avoided by the use of a common online platform for health-care staff and patients. This platform could be utilized for medical communication as well as transfer of medical records (32). Regulations should outline requirements in terms of handling of sensitive patient data and risk communication (33).

Additionally, legal frameworks focused on the provision of social support need to be considered both during the treatment and follow-up processes. Social support can decrease stress, improve patients' well-being while traveling, and improve the recovery process (34, 35). With this in mind, the definition of juridical facilitation of travel arrangements could be of great benefit to medical travelers. This could entail the provision of subsidies for costs of an accompanying significant other, as well as less stringent visa procedures, when accompanying a relative during the medical travel process.

CONCLUDING REMARKS

The implementation of a GHAP is crucial to ensure safe and economically beneficial medical travel for a broader population. In inbound countries, the promotion of medical travel has the capacity to counter increasing costs of health care by creating new markets and increased revenue for health services, and lead to more efficient use of resources and labor benefiting local populations economies, and health-care systems. In outbound countries, medical travel can provide access to treatment that is more affordable or unavailable domestically. Further, legal consequences must be considered at each phase of the medical travel process. Legal considerations include the standardization of visa procedures, the implementation of international quality standards, the accreditation of health-care providers and third-party agencies, the protection of the local population and the monitoring of information provided to patients to ensure informed decision making for the patients and the families involved. Fairness, equal access to treatments for medical travelers and local populations, equity, and long-term sustainability are key components when developing a GHAP framework. Considering the existing risk for all stakeholders, it is essential that existing economic and legal barriers to medical travel are accounted for in the establishment of a comprehensive GHAP.

AUTHOR CONTRIBUTIONS

All authors contributed to the data collection for the paper. NR, JT, SB, OK, KR, and SHK contributed to the writing of the paper. In addition, KR, CD, and SHK contributed to the design of the study.

CONFLICT OF INTEREST STATEMENT

The authors declare that the research was conducted in the absence of any commercial or financial relationships that could be construed as a potential conflict of interest.

ACKNOWLEDGMENTS

We are thankful to Corpus Christi College for their support of the Junior Researcher Programme, which provided the framework for this project. We would also like to thank Ebba Karlsson, Anna Müller, and Daisy Mechelmans for their contributions to this work.

REFERENCES

1. Lunt N, Smith R, Exworthy M, Green ST, Horsfall D, Mannion R. Medical Tourism: Treatments, Markets and Health System Implications: A Scoping Review. Paris: OECD; (2011).

2. Ruggeri K, Záliš L, Meurice CR, Hilton I, Ly T, Zupan Z, et al. Evidence on global medical travel. Bull World Health Organ (2015) 93:785–9.

3. Crozier GKD, Baylis F.. The ethical physician encounters international medical travel. J Med Ethics(2010) 36:297–301.

4. Keckley PH, Eselius LL. 2009 Survey of Health Care Consumers: Key Findings, Strategic Implications. Washington: Deloitte Center for Health Solutions; (2009).

5. Herrick DM. Medical Tourism: Global Competition in Health Care. Dallas, TX: National Center for Policy Analysis; (2007).

6. Cheung IK, Wilson A.. Arthroplasty tourism. Med J Aust (2007) 187:666–7.

7. Chen LH, Wilson ME.. The globalization of healthcare: implications of medical tourism for the infectious disease clinician. Clin Infect Dis (2013) 57:1752–9.

8. European Parliament. Directive 2011/24/EU of the European Parliament and of the Council of 9 March 2011 on the Application of Patients' Rights in Cross-Border Healthcare (2011). Available from: http://eur-lex.europa.eu/legal-content/EN/TXT/?uri=CELEX:32011L0024

9. Crooks VA, Turner L, Cohen IG, Bristeir J, Snyder J, Casey V, et al. Ethical and legal implications of the risks of medical tourism

for patients: a qualitative study of Canadian health and safety representatives' perspectives. BMJ Open (2013) 3:e002302.

10. Cormany D, Baloglu S. Medical travel facilitator websites: an exploratory study of web page contents and services offered to the prospective medical tourist. Tour Manag (2011) 32:709–16.10.1016/j.tourman.2010.02.008

11. Baker D. Globalising Healthcare: A Prescription with Benefits. Washington, DC: OECD; (2010). Observer No. 281.

12. Whittaker A, Manderson L, Cartwright E.. Patients without borders: understanding medical travel. Med Anthropol (2010) 29:336–43.

13. Karp L. Medical Travel: Global Impact and Local Response. Thomas Jefferson University Health Policy Newsletter 21; (2008). p. 1–4.

14. Gifford B, Park S, Anand S.. Perceptions about the impact of global medical travel on poorer populations in India. World Hosp Health Serv (2009) 45(4):6–10.

15. Helble M.. The movement of patients across borders: challenges and opportunities for public health. Bull World Health Organ (2011) 89:68–72.10.2471/BLT.

16. Lee CG. Heath care and tourism: evidence from Singapore. Tour Manag (2010) 31:486–8.

17. Lee CG, Hung WT. Tourism, health and income in Singapore. Int J Tour Res (2010) 12:355–9.

18. Chanda R.. Trade in health services. Bull World Health Organ (2002) 80:158–63.

19. Wibulpolprasert S, Pachanee C, Pitayarangsarit S, Hempisut P.. International service trade and its implications for human resources for health: a case study of Thailand. Hum Resour Health (2004)2:

20. Lunt NT, Mannion R, Exworthy M. A framework for exploring the policy implications of UK medical tourism and international patient flows. Soc Policy Adm (2013) 47:1–25

21. Turner L. "First world health care at third world prices": globalization, bioethics and medical tourism.Biosocieties (2007) 2:303–25.

22. Mohamad NW, Omar A, Haron SM. The moderating effect of medical travel facilitators in medical tourism. Procedia Soc Behav Sci (2012) 65:358–63.

23. Turner LG.. Quality in health care and globalization of health services: accreditation and regulatory oversight of medical tourism companies. Int J Qual Health Care (2011) 23:1–7.

24. Canales MT, Kasiske BL, Rosenberg ME.. Transplant tourism: outcomes of United States residents who undergo kidney transplantation overseas. Transplantation (2006) 82:1658–61.

25. Panteli D, Wagner C, Verheyen F, Busse R.. Continuity of care in the cross-border context: insights from a survey of German patients treated abroad. Eur J Public Health (2015) 25:557–63.

26. Terzi E, Kern T, Kohnen T. Complications after refractive surgery abroad. Ophthalmologe (2008)105:474–9.

27. Kiasuwa Mbengi RL, Baeten R, McKee M, Knai C.. Issues arising when crossing a border to give birth: an exploratory study on the French-Belgian border. Facts Views Vis Obgyn (2014) 6:127–32.

28. Glonti K, Hawkesworth S, Footman K, Doering N, Schmidt AE, Destrebeq F, et al. European health professionals' experience of cross-border care through the lens of three common conditions. Eur J Integr Med(2015) 7:29–35.

29. Legido-Quigley H, McKee M, Nolte E, Glinos IA. Assuring the Quality of Health Care in the European Union. World Health Organization; (2008). Available from: www.euro.who.int/document/e91397.pdf

30. Kácha O, Kovács BE, McCarthy C, Schuurmans AAT, Dobyns C, Haller E, et al. An approach to establishing international quality standards for medical travel. Front Public Health (2016)4:29.

31. Arce HE. Hospital accreditation as a means of achieving international quality standards in health. Int J Qual Health Care (1998) 10:469–72.

32. Santoro A, Silenzi A, Ricciardi W, McKee M.. Obtaining health care in another European Union Member State: how easy is it to find relevant information? Eur J Public Health (2015) 25:29–31.

33. Jovanovic B. Hospital accreditation as method for assessing quality in healthcare. Arch Oncol (2005)13:156–7.

34. Uchino BN.. Social support and health: a review of physiological processes potentially underlying links to disease outcomes. J Behav Med (2006) 29:377–87.

35. Thoits PA.. Mechanisms linking social ties and support to physical and mental health. J Health Soc Behav (2011) 52:145–61.

Index